2006
x 3|09 · 3|10
3+5|10 10|11

Other Books by the Author

The Complete Book of Locks and Locksmithing, Sixth Edition
Locksmithing
The Complete Book of Electronic Security

The Complete Book of Home, Site, and Office Security

Selecting, Installing, and Troubleshooting Systems and Devices

Bill Phillips

3 1336 07461 3846

McGraw-Hill

New York   Chicago   San Francisco   Lisbon   London   Madrid
Mexico City   Milan   New Delhi   San Juan   Seoul
Singapore   Sydney   Toronto

The McGraw·Hill Companies

Library of Congress Cataloging-in-Publication Data

Phillips, Bill, date.
  The complete book of home, site, and office security: selecting, installing, and troubleshooting systems and devices / Bill Phillips.
     p.   cm.
  Includes index.
  ISBN 0-07-146744-0 (alk. paper)
  1. Security systems. 2. Dwellings—Security measures. I. Title.
  TH9705.P485 2006
  658.4′7—dc22

                            2006015534

1 2 3 4 5 6 7 8 9 0 DOC/DOC 0 1 3 2 1 0 9 8 7 6

ISBN 0-07-146744-0

The sponsoring editor for this book was Cary Sullivan, the editing supervisor was David E. Fogarty, and the production supervisor was Richard C. Ruzycka. It was set in Century Schoolbook by International Typesetting and Composition. The art director for the cover was Anthony Landi.

Printed and bound by RR Donnelley.

This book was printed on acid-free paper.

McGraw-Hill books are available at special quantity discounts to use as premiums and sales promotions, or for use in corporate training programs. For more information, please write to the Director of Special Sales, McGraw-Hill Professional, Two Penn Plaza, New York, NY 10121-2298. Or contact your local bookstore.

To my sister and best friend Merlyn Smith-Coles

Contents

Preface

For about the last 20 years I've been writing books and articles about various aspects of physical and electronic security, including *The Complete Book of Locks and Locksmithing*, 6th edition, and *The Complete Book of Electronic Security*, both published by McGraw-Hill. But I haven't written a comprehensive book on home, office, and site security since 1994. A lot has changed since then, so I decided to write this book.

Whether you want to become a better security professional, improve the security of your own home, office, or site, or hire a security professional, this book can help you. You can use it to prepare for a variety of professional security certification and registration tests, including the certified protection professional (CPP), the registered professional locksmith (RPL), and the registered security professional (RSP). Copies of the RPL and RSP tests are included in Appendixes A and B, respectively.

This book covers a wide range of safety- and security-related topics, including strengthening doors and windows, choosing the best locks, using state-of-the-art access control systems, lighting for safety and security, using closed-circuit television systems, improving computer security, defending against terrorism, fire safety, and providing security consulting.

This book will teach you how to think like a security professional and not like a fire victim or burglar. That's because most burglars don't know a whole lot about security. They generally just take advantage of easy opportunities because most people aren't very security conscious.

You've probably heard these three myths:

1. If burglars want to get in badly enough, you can't keep them out.

2. The only way to keep burglars out is to use expensive high-tech devices.

3. Most home and office fires are unavoidable without taking expensive precautions.

Often such myths are perpetuated in books and articles written by well-meaning journalists who use crime or fire victims or "former burglars" as primary sources of information. But neither a burglar nor a crime victim is a security expert.

Contrary to movies and television crime dramas, the criminals who break into homes are not James Bond clones. Most home burglars know little about defeating locks or electronic security systems. Many rely on just one or two quick, quiet, and simple entry techniques, and they search for homes at which they can use those methods.

Home security articles and books that are based on advice from burglars often contain misleading, incomplete, and potentially dangerous information. In many cases the advice one burglar gives to thwart his or her favorite break-in technique can make it easier for other burglars to use their techniques.

Good security advice isn't based on scattered facts or personal anecdotes but rather on a comprehensive understanding of security measures, security hardware, and electronic systems. I've never been burglarized, attacked, or had a car stolen, nor have I ever been a criminal, but I have had years of training and experience in locksmithing, safe and vault work, alarm system installation, martial arts, and security consulting. I've also lived in some of the highest crime areas in the United States, including New York City and Detroit. I know that it is possible for people to stay safe and protect their property without spending a lot of money.

As you read this book, you'll learn how to assess intelligently safety and security needs and how best to meet those needs in a cost-effective way. If you or someone you know has been victimized by household crime or fire, you'll probably discover within these pages why the incident occurred and how it might have been prevented.

Each chapter includes do-it-yourself projects, money-saving tips, and practical advice that you can use right now, whether you live in a large city or a small town. Nearly every sentence of this book is written in nontechnical language, so you should have no trouble following the advice and instructions. However, I've included a Glossary that should be helpful if you find an unfamiliar word.

Chapter 16 includes a comprehensive home safety and security checklist to guide you in surveying a home. You may want to glance at it now, but you'll benefit more by using the checklist after you've read the whole book.

Because all the topics in this book are interrelated, you'll get the most out of it by reading it from beginning to end. Even if you skip around, you'll discover new ways to protect your home, your car, your family, and your business.

If you have questions or comments about this book, I'd love to hear from you. You can write to me at: Box 2044, Erie, PA 16512-2044, or e-mail me at LocksmithWriter@aol.com.

BILL PHILLIPS

Acknowledgments

I want to thank the following companies for helping me with this book: Adams Rite Manufacturing, Aiphone, Baldwin Hardware Corporation, Black & Decker Corporation, Black Widow Security Systems, eKey USA Systems, Gardall Safe Corporation, HighPower Security Products, LLC, Intermatic, Inc., Kaba Access Control, M.A.G. Manufacturing, Master Lock Company, Medeco Security Locks, Inc., Mul-T-Lock USA, Inc., NAPCO Security Systems, Inc., Progress Lighting, REDDCO, Inc., Securitron Magnalock Corp., and Visonic, Inc. Special thanks to my McGraw-Hill editor, Cary Sullivan, and to Patricia Bruce and Jonathan Gavin.

The Complete Book of
Home, Site, and
Office Security

History of Loss Prevention

Broadly speaking, *security* refers to the military, public law enforcement, and private security. This book focuses on the private security sector. Private security covers such things as the use of locks, safes, alarms, security officers, computer security, fire safety, insurance, and securing hazardous materials. Because of the broad range of private security issues, many security professionals prefer to use the term *loss prevention* or *assets protection* rather than *security*.

Saul D. Astor pointed out

> In the minds of many, the very word "security" is it's own impediment. . . . Security carries a stigma; the very word suggests police, badges, alarms, thieves, burglars, and some generally negative and even repellent mental images. . . . Simply using the term "loss prevention" instead of the word "security" can be a giant step toward improving the security image, broadening the scope of the security function, and attracting able people.

Loss prevention refers to all methods of reducing thefts, injuries, and deaths, including not only locks, alarms, and safes but also fire safety, computer security, insurance, and accident prevention. For those reasons, this book refers to *loss-prevention officers* rather than *security officers*.

Early Civilizations

The earliest forms of loss prevention were physical barriers. Such barriers, including fences and walls, are still important for security today. Early human beings relied on caves for protection from enemies and wild animals. They would roll a large rock in the front for protection. In warm climates, trees and thickets were used. Prehistoric Pueblo Indians used ladders to get to their cliff dwellings and would pull the ladders in at night. This wasn't foolproof, though, because enemies would bring their own ladders. Another early security strategy was to build homes on sunken pilings or platforms on lakes. Access to the dwellings required a boat or a drawbridge.

The longest security structure ever built is the Great Wall of China, which was built over hundreds of years beginning around 400 B.C.E. Thousands of workers lived near the wall and helped create the 4,000-mile structure that reached heights of 25 feet. In addition, 40-foot-tall watchtowers were located every 100 to 200 yards. The Chinese built the wall to protect their northern border from invaders. The wall only protected against minor attacks, however. It was little help against a major attack. Genghis Khan, for instance, swept across the wall during the 1200s and conquered much of China. The wall is mostly collapsed now, but in 1949, the Chinese government restored some of it, and it is now a major tourist attraction.

As societies became more complex, inequities occurred, and wealthy people needed more protection from thieves. Part of such protection included severe punishments, such as branding, burning, mutilating, and crucifying. A person's body showed his criminal record. By 1750 B.C.E., the laws of Hammurabi, King of Babylon, codified the responsibilities of the individual and rules for dealings between individuals, as well as retribution for breaking the rules. These laws are engraved on an 8-foot pillar and are the earliest record of laws to protect people and property.

The earliest locks may no longer be around, and there may be no written records of them. How likely it is for old locks to be found depends on the materials they were made from and on the climate and various geologic conditions they have been subjected to over the years. There is evidence to suggest that different civilizations probably developed the lock independently of each other. The ancient Egyptians, Romans, and Greeks are credited with inventing the oldest known types of locks.

The oldest known lock was found in 1842 in the ruins of Emperor Sargon II's palace in Khorsabad, Persia. The ancient Egyptian lock was dated to be about 4,000 years old. It relied on the same pin-tumbler principle that is used by many of today's popular locks.

The Egyptian lock consisted of three basic parts: a wood crossbeam, a vertical beam with tumblers, and a large wood key. The crossbeam ran horizontally across the inside of the door and was held in place by two vertically mounted wooden staples. Part of the length of the crossbeam was hollowed out, and the vertical beam intersected it along the hollowed-out side. The vertical beam contained metal tumblers that locked the two pieces of wood together. Near the tumbler edge of the door there was a hole accessible from outside the door that was large enough for someone to insert the key and an arm. The spoon-shaped key was about 14 inches to 2 feet long with pegs sticking out of one end. After the key was inserted into the keyhole (or *armhole*), it was pushed into the hollowed-out part of the crossbeam and vertical beam so that the pins no longer obstructed movement of the crossbeam. Then the crossbeam (or *bolt*) could be pulled into the open position. To see how the lock worked, see Figure 1.1.

Cutaway view of vertical beam with tumblers

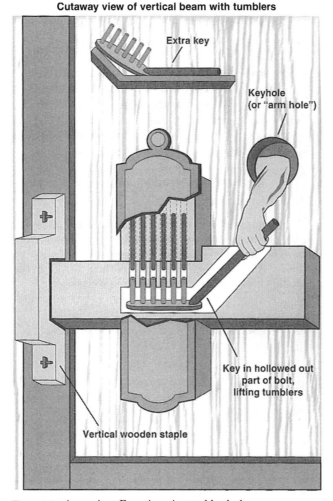

Figure 1.1   An ancient Egyptian pin-tumbler lock.

Ancient Greece

Ancient Greece became advanced commercially and culturally rich between the ninth and the third centuries B.C.E. As civilization advanced, so did the need for protection. The Greeks protected themselves with the use of the *polis*, or "city-state," which consisted of a city and the surrounding land protected by a central fortress built on a hill. A hierarchical society made the ruling class fear revolts from poor people. Spartans kept spies among the poor people and subversives. The first police evolved to protect local communities during the time of the Greek city-states. Citizens were responsible for policing their communities.

The Greek rulers didn't consider policing to be a state responsibility. When problems arose, they used the army. During this era, the Greek philosopher Plato introduced a new concept of justice in which the criminal not only would be punished but also would be required to take some form of rehabilitation.

Most early Greek doors pivoted at the center and were secured with rope tied into intricate knots. The cleverly tied knots, along with beliefs about being cursed for tampering with them, provided some security. When more security was needed, doors were secured with bolts from the inside. In the few cases where locks were used, they were primitive and easy to defeat. The Greek locks used a notched boltwork and were operated by inserting the blade of an iron sickle-shaped key, about a foot long, in a key slot and twisting it 180 degrees to work the bolt (Figure 1.2). These locks could be defeated just by trying a few different sized keys.

In about 800 B.C.E., the Greek poet Homer described the Greek lock in his poem *The Odysseus*:

> She went upstairs and got the store room key, which was made of bronze and had a handle of ivory; she then went with her maidens into the store room at the end of the house, where her husband's treasures of gold, bronze, and wrought iron were kept. . . . She loosed the strap from the handle of the door, put in the key, and drove it straight home to shoot back the bolts that held the doors.

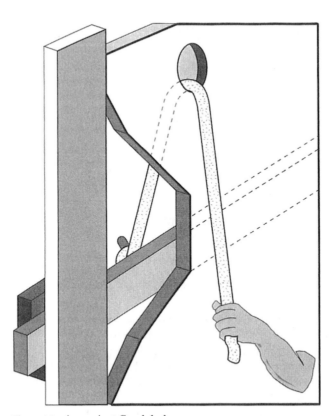

Figure 1.2 An ancient Greek lock.

Ancient Rome

Before the birth of Christ, the civilization of Rome developed both commercially and culturally. Rome was located just 15 miles from the sea and easily could share in the trade of the Mediterranean. The city sat on seven hills overlooking the Tiber River, which allowed ease in fortification and defense. Geese were kept at strategic locations so that they would squawk when an army approached. This was an early intruder alarm.

The chief business of the Roman state was war, which it was well capable of carrying out. A body of 8,000 troops equipped with helmets, shields, lances, and swords formed the basic unit of a Roman army. Later, a more maneuverable unit of 3,600 soldiers armed with iron-tipped javelins was used. These men also were used to keep law and order. During the time of Christ, the first emperor of Rome, Caesar Augustus (27 B.C.–A.D. 14), established the Praetorian Guard to protect him and his property. From this force emerged the urban cohorts of 500 to 600 men to protect the peace of the city. Many security professionals believe that the guard was the most effective police force until recent developments in law enforcements. Modern-day coordinated patrolling and crime-prevention methods began with the subsequent nonmilitary vigils. They were night watchmen who also served as firefighters. Rome also created the Vigilantes, which was a secret unit to protect the state.

Like the Greeks, the early Romans used locks with notched boltworks. The Romans improved on the lock design in many ways, however, such as by putting the boltwork in an iron case and using keys of iron or bronze. Because iron rusts and corrodes, few early Roman locks still exist. But a lot of the keys are around. Often the keys were ornately designed to work as jewelry, either as finger rings or as necklaces using string.

Two of the most important innovations of the Roman locks were the spring-loaded bolt and the use of wards on the case. The extensive commerce during the time of Julius Caesar led to a great demand for locks among wealthy merchants and politicians. The type of lock used by the Romans, the warded bit-key lock, is still being used today in many older homes. Because the lock provides little security, typically it's found on interior doors, such as closets and sometimes bathrooms.

The Romans sometimes are credited with inventing the padlock, but that's controversial. There is evidence that the Chinese may have invented it independently before or at about the same time.

The demand for locks declined after the fall of Rome in the fifth century because people had little to protect. The few locks used during the period were specially ordered for nobility and the few wealthy merchants.

Europe in the Middle Age

It's important to understand England's security heritage because much of America's customs, language, laws, locks, and security methods can be traced to its English heritage. The Middle Ages involve the period of history between the fifth and fifteenth centuries and constitute a line of demarcation between ancient and modern times.

Feudalism developed in Europe during the Dark Ages, after destruction of the ancient Greek and Roman empires. Overlords provided food and security to farmers and protected castles by fortified walls, towers, and a drawbridge that could be raised across a moat. At that time, security required registration, licensing, and a fee. King Henry II of England (reigned from A.D. 1154–1189) destroyed over 1,100 unlicensed castles that had been built during a civil war.

The war band of the early Germans, the *comitatus*, was a mutual arrangement in which a leader commanded the loyalty of followers to fight together and win booty. (Today, the term *posse comitatus* refers to a body of citizens that the authority can call on for assistance against criminals.) Many land owners created private armies to defend against bands of German barbarians.

The concept of tithing and the frankpledge system resulted in increased protection between the seventh and tenth centuries. The frankpledge system, which started in France and spread to England, emphasized communal responsibility for justice and protection. The tithing, or group of ten families, shared the duties of protecting the community and maintaining the peace.

In 1066, William, Duke of Normandy, crossed the English Channel and defeated the Anglo-Saxons at Hastings. Under martial law, a highly repressive police system developed as the state appropriated responsibility for peace and protection. The tithing system and community authority were weakened. William divided England into 55 districts, or *shires*. A *reeve*, from the military, was assigned to each district. (The word *sheriff* was derived from *shire-reeve*.) William is credited with changing the law to make theft a crime against the state rather than against an individual and was instrumental in separating police functions from judicial functions. Defendants who were arrested by the shire-reeves were tried by a traveling judge.

In 1215, King John signed the Magna Carta, which guaranteed civil and political liberties. Local government power increased at the expense of the national government, and at the local level, community protection increased.

The Statute of Westminster (1285) was another security milestone issued by King Edward I to organize a police and justice system. Every town was required to send out men at night to close the gates of walled towns and to enforce curfews.

During the Middle Ages, metal workers in England, Germany, and France continued to make warded locks, with no significant security changes. They focused on making elaborate, ornately designed cases and keys. Locks became works of art.

In the fourteenth century, the locksmiths' guild came into prominence. The guild required journeymen locksmiths to create and submit a working lock and key to the guild before being accepted as master locksmiths. The locks and keys weren't made to be installed but to be displayed in the guildhall. The guild's work resulted in some beautiful locks and keys. The problem with the locksmiths' guild is that it gained too much control over locksmiths and didn't require technological advances. Because of the locksmiths' guilds, few significant security innovations were made. The major innovations included such things as false and hidden keyholes. A fish-shaped lock, for instance, might have a keyhole hidden behind a fin.

Keys were made that could move about a post and shift the position of a movable bar (the locking bolt). The first obstacles to unauthorized use of the lock were internal wards. Medieval and renaissance artisans improved on the warded lock by using many interlocking wards and more complicated keys. But many of the wards could be bypassed easily.

In France in 1767, the treatise *The Art of the Locksmith* was published; it described examples of the lever-tumbler lock. The inventor of this lock is unknown. As locksmithing advanced, locks were designed with multiple levers, each of which had to be lifted and aligned properly before the bolt could move to the unlocked position.

The Eighteenth Century

The Industrial Revolution, which occurred during the eighteenth century, resulted in many people coming into the cities to work. Crime spiraled in large cities such as London. Citizens carried arms and hired private security to protect themselves. The military was called to quell riots when they occurred.

Several individuals were instrumental in reducing crime. In 1749, for instance, Henry Felding published the social novel *Tom Jones*. The book depicted the troubles in the London slums. Felding also worked for social reform and created a strategy for using police to prevent crime. He suggested a foot patrol to secure the streets, a mounted patrol for highways, and the "Bow Street Runners"— which are credited with being the first detective unit. The Bow Street Runners moved quickly to investigate crimes. The merchant police were formed to protect businesses, and the Thames River police provided protection at the docks. During this time, more than 160 crimes, including stealing food, were punishable by death.

Even with all these security forces, however, crime continued to spiral as cities created more and more slums. Citizens protected themselves with firearms and by placing wolf traps near doors and windows.

Little progress was made in lock security until the eighteenth century. At that time, cash awards and honors were given as incentives to those who could pick open new and more complex locks. This resulted in more secure lock designs. In the forefront of lock design were three Englishmen: Robert Barron, Joseph Bramah, and Jeremiah Chubb.

Robert Barron patented the first major improvement over warded locks. He added the tumbler principle to wards for increased security. His double-acting lever-tumbler lock was more secure than other locks during that time and remains today the basic design for lever-tumbler locks. As with other lever-tumbler locks, Barron's used wards. But he also used a series of lever tumblers,

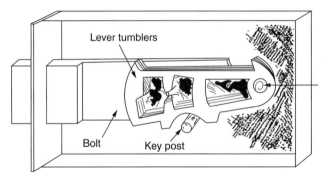

Figure 1.3   An early lever-tumbler lock.

each of which was acted on by a separate step of the key. If any tumbler wasn't raised to the right height by the key, its contact with a bolt stump would obstruct bolt movement. Barron's lock corrected the shortcomings of earlier lever-tumbler locks, which could be circumvented easily by any key or instrument thin enough to bypass the wards. Barron added up to six of these double-lever actions to his lock and thought it was virtually impossible to open it except with the proper key (Figure 1.3). He soon found out differently.

Another Englishman, Joseph Bramah, wrote *A Dissertation on the Construction of Locks*, which exposed the many weaknesses of existing so-called thief-proof locks and pointed out that many of them could be picked by a good specialist or criminal with some training in locks and keys. Bramah admitted that Barron's lock had many good points but also revealed its major fault: The levers, when in the locked position, gave away the lock's secret. The levers had uneven edges at the bottom; thus a key coated with wax could be inserted in the lock, and a new key could be made by filing where the wax had been pressed down or scraped away. Several tries could create a key that matched the lock. Bramah pointed out that the bottom edges of the levers showed exactly the depths to which the new key should be cut to clear the bolt. Bramah suggested that the lever bottoms should be cut unevenly. Then only a master locksmith should be able to open it.

Using those guidelines, Bramah patented a barrel-shaped lock in 1798 (Figure 1.4) that employed multiple sliders around the lock that were to be aligned with corresponding notches around the barrel of its key. The notches on the key were of varying heights. When the right key was pushed into the lock, all the notches lined up with the sliders, allowing the barrel to rotate to the unlocked position. It was the first lock to use the rotating element within the lock itself.

During this period, burglary was a major problem. After the Portsmouth, England, dockyard was burglarized in 1817, the British Crown offered a reward to anyone who could make an unpickable lock. A year later, Jeremiah Chubb patented his lock and won the prize money.

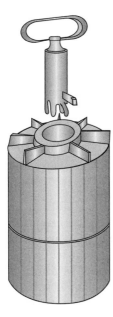

Figure 1.4 A Bramah barrel-shaped lock.

Chubb's detector lock was a four-lever tumbler-rim lock that used a barrel key. It had many improvements over Barron's lock. One of the improvements was a metal "curtain" that fell across the keyhole when the mechanism began to turn, making the lock hard to pick. Chubb's lock also added a detector lever that indicated that the lock had been tampered with. A pick or an improperly cut key would raise one of the levers too high for the bolt gate. That movement engaged a pin that locked the detector lever. The lever could be cleared by turning the correct key backward and then forward.

Chubb's lock got much attention. It was recorded that a convict who had been a lock maker was on board one of the prison ships at Portsmouth dockyard and said that he had easily picked open some of the best locks and that he could easily pick open Chubb's detector lock. He was given one of the locks and all the tools that he asked for, including key blanks fitted to the drill pin of the lock. As incentive to pick open the lock, Mr. Chubb offered the convict a reward of £100, and the government offered a free pardon if he succeeded. After trying for several months to pick the lock, he gave up. He said that Chubb's lock was the most secure lock he had ever met with and that it was impossible for anyone to pick it with false instruments. The lock was improved by Jeremiah's brother, Charles Chubb, and Charles' son, John Chubb, in several ways, including the addition of two levers and false notches on the levers.

The lock was considered unpickable until it was picked open in 1851 at the International Industrial Exibition in London by an American locksmith named Alfred C. Hobbs. At that event, Hobbs picked open both the Bramah and the Chubb locks in less than an hour.

Contemporary England

Repeated attempts were made over the next 500 years to improve protection and justice in England. Each king faced increasingly serious crime problems and cries from the citizenry for help. As England colonized many parts of the world, and as trade and commercial pursuits brought many people into the cities, urban problems and high crime rates continued. Unsatisfied by the protection they were receiving, merchants hired private security forces to protect their businesses.

Peel's Reforms

The Metropolitan Police Act came about through the efforts of Sir Robert Peel in 1829, which was the birth of modern policing. His innovative ideas were accepted by Parliament, and he was chosen to implement an act that established a full-time, unarmed police force with the main purpose of patrolling London. Peel is also credited with reforming the criminal law by limiting its scope and abolishing the death penalty for more than 100 crimes. It was hoped that such reforms would gain public support and respect for the police. Peel was selective in his choice of hiring, and he emphasized professional training. His reforms are applicable today and include crime prevention, the strategic deployment of police according to time and location, record keeping, and distribution of crime news.

The Growth of Policing

During the mid 1800s, there was a turning point in both law enforcement and private security in England and America. Several major cities, including New York, Philadelphia, and San Francisco, organized police forces—many of which were modeled after the London Metropolitan Police. But corruption was widespread. Many urban police forces received large boosts in personnel and resources to combat the growing militancy of the labor unions in the late 1800s and early 1900s. Many of the large urban police forces were formed originally as strikebreakers. In 1864, the U.S. Treasury had already established an investigative unit. As in England, an increase in paid police officers didn't eliminate the need for private security.

Early America

The Europeans who colonized North America brought the legal and security heritage of their mother countries. The watchman system and collective responses were popular. A central fortification in largely populated areas provided increased security from crime. As communities grew, the office of sheriff took hold in the South, whereas the functions of constable and watchman were the norm in the Northeast. The sheriff's duties included catching criminals, serving subpoenas, and collecting taxes. Constables were responsible for keeping the peace, bringing suspects and witnesses to court, and eliminating health hazards. As in England, the watch system was inefficient, and those convicted of minor crimes were sentenced to serve time on the watch.

During America's early years, England had a policy against its skilled artisans leaving the country. This was to keep them from running off and starting competing foreign companies. Locks made by early American locksmiths didn't sell well. In the mid 1700s, few colonists used door locks, and most that were used were copies of European models. More often, Americans used lock bolts mounted on the inside of the door that could be opened from the outside by a latchstring, hence the phrase, "the latchstring's always out." At night, the string would be pulled inside, "locking" the door. Of course, someone had to be inside to release the bolt. An empty house was left unlocked. As the country settled, industry progressed and theft increased, increasing the demand for more and better locks. American locksmiths soon greatly improved on the English locks and were making some of the most innovative locks in the world. Before 1920, American lock makers patented about 3000 different locking devices.

In 1805, an American physician, Abraham O. Stansbury, was granted an English patent for a pin-tumbler lock that was based on the principles of both the Egyptian and Bramah locks. Two years later, the design was granted the first lock patent by the U.S. Patent and Trademark Office. Stansbury's lock used segmented pins that automatically relocked when any tumbler was pushed too far. The double-acting pin-tumbler lock was never manufactured for sale.

In 1836, a New Jersey locksmith, Solomon Andrews, developed a lock that had adjustable tumblers and keys, which allowed the owner to rekey the lock anytime. Because the key also could be modified, there was no need to use a new key to operate a rekeyed lock. But few homeowners used the lock because rekeying it required dexterity, practice, and skill. The lock was of more interest to banks and businesses.

In the 1850s, two inventors, Solomon Andrews and Robert Newell, were granted patents on an important new feature—removable tumblers that could be disassembled and scrambled. The keys had interchangeable bits that matched the various tumbler arrangements. After locking up for the night, a prudent owner would scramble the key bits. Even if a thief got possession of the key, it would take hours to stumble onto the right combination. In addition to removable tumblers, this lock featured a double set of internal levers.

Newell was so proud of this lock that he offered a reward of $500 to anyone who could open it. A master mechanic took him up on the offer and collected the money. This experience convinced Newell that the only secure lock would have its internals sealed off from view. Ultimately, sealed locks appeared on bank safes in the form of combination locks.

Until the time of A. C. Hobbs, who picked the famed English locks with ease, locks were opened by making a series of false keys. If the series was complete, one of the false keys would match the original. Of course, this procedure took time. Thousands of hours might pass before the right combination was found. Hobbs depended on manual dexterity. He applied pressure on the bolt while manipulating one lever at a time with a small pick inserted through the keyhole. As each lever tumbler unlatched, the bolt moved a hundredth of an inch or so.

Until the early nineteenth century, locks were made by hand. Each locksmith had his own ideas about the type of mechanism—the number of lever tumblers, wards, and internal cams to put into a given lock. Keys contained the same individuality. A lock could have 20 levers and weigh as much as 5 pounds.

In 1833, three brothers, Eli Whitney Blake, Philo Blake, and John Blake—the Blakebrothers—were granted a patent for a unique door latch that had two connecting doorknobs. It was installed by boring two connecting holes. The larger hole, which was drilled through the door face, was for the knob mechanism. The smaller hole, which was drilled through the door edge, was for the latch. The big difference between their latch and others of their time was that all door locks were installed by being surface mounted to the inside surface of a door. In 1834, the brothers formed the Blake Brothers Lock Company to produce and sell their unusual latch. At that time, the brothers probably never imagined that nearly 100 years later their development would be used to revolutionize lock designs.

In 1844, Linus Yale, Sr., of Middletown, Connecticut, patented his Quadruplex bank lock, which incorporated a combination of ancient Egyptian design features and mechanical principles of the Bramah and Stansbury locks. The Quadruplex had a cylinder subassembly that denied access to the lock bolt. In 1848, Yale patented another pin-tumbler design based on the Egyptian and Bramah locks. His early models had the tumblers built into the case of the lock and had a round fluted key. His son, Linus Yale, Jr., improved on the lock design and is credited with inventing the modern pin-tumbler lock.

Arguably, the most important modern lock development is the Yale Mortise Cylinder Lock, U.S. Patent No. 48,475, issued on June 27, 1865, to Linus Yale, Jr. It turned the lock-making industry upside down and established a new standard. Yale, Jr.'s, lock not only could be rekeyed easily, but it also provided a high level of security, could be mass produced easily, and could be used on doors of various thicknesses. His lock design meant that keys no longer had to pass through the thickness of the door to reach the tumblers or bolt mechanism, which allowed the keys to be made thinner and smaller. (Linus Yale, Jr.'s, first pin-tumbler locks used a flat steel key rather than the paracentric cylinder-type key often used today.)

Since 1865, there have been few major changes to the basic design of mechanical lock cylinders. Most cylinder refinements since that time have been limited to using unique keyways (along with correspondingly shaped keys), adding tumblers, varying tumbler positions, varying tumbler sizes and shapes, and combining two or more basic types of internal construction—such as the use of both pin tumblers and wards. Most major changes in lock design have centered around the shape and installation methods of the lock.

In 1916, Samuel Segal, a former New York City police officer, invented the jimmy-proof rim lock (or interlocking deadbolt). The surface-mounted lock has

Figure 1.5 An interlocking deadbolt. (*Courtesy of M.A.G. Manufacturing.*)

vertical bolts that interlock with "eye-loops" of its strike, locking the two parts together in such a way that you would have to break the lock to pry them apart (Figure 1.5).

In 1920, Frank E. Best received his first patent for an interchangeable-core lock. This lock allows you to rekey it just by using a control key and switching cores. The core was made to fit into padlocks, mortise cylinders, deadbolts, key-in-knobs, and other types of locks (Figure 1.6).

Figure 1.6 An interchangeable core cylinder. (*Courtesy of Medeco High Security Locks.*)

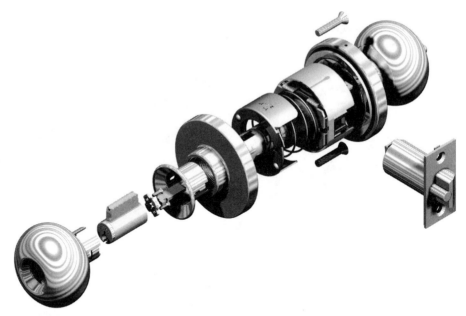

Figure 1.7 An exploded view of a modern key-in-knob lock. (*Courtesy of Medeco High Security Locks.*)

In 1928, Walter Schlage patented a cylindrical lock that incorporates a locking mechanism between the two knobs. Schlage's was the first knob-type lock to have mass appeal. Today, key-in-knob locks are commonplace. Figure 1.7 shows a modern key-in-knob lock.

In 1933, Chicago Lock Company introduced its tubular-key lock, called the Chicago Ace Lock. The lock was based on the pin-tumbler principle but used a circular keyway. The odd keyway made it hard to pick open without using special tools. For a long time, many locksmiths referred to all tubular key locks as "Ace locks," not realizing that it was a brand name. Today, the lock is made by many manufacturers and is used on vending machines and in padlocks and bicycle locks.

In 1832, the English lock maker Stephen G. Bucknall became the first trunk and cabinet lock manufacturer in America. He made about 100 cabinet lock models but didn't sell very many. After his company folded, Bucknall went to work for Lewis, McKee & Company. William E. McKee was a major investor in the company. In 1835, Bucknall left the company and received financial assistance from McKee to form the first trunk lock company in America, called the Bucknall, McKee Company. A couple of years later, Bucknall sold the business and went to work for North & Stanley Company.

One of the greatest successes and failures of the American lock industry was the Eagle Lock Company, formed in 1854. It was the result of a merger between the James Terry Company and Lewis Lock Company. Eagle Lock had money to burn and was quick to buy out its competitors, such as American Lock Company, Gaylord, and Eccentric Lock Company. In 1922, the company had 1800 employees and several large warehouses. In 1961, the company introduced a popular line of locks and cylinders called Super-Security. They were highly pick-resistant, and their keys couldn't be duplicated on standard key machines. The company was sold to people who didn't know a lot about locks, and made bad decisions that sent the company on a downward path. Profits were being siphoned off without considering the long-term needs of the business. The top management quit. By the early 1970s, the company was barely holding on. In 1973, a businessperson bought it from Penn-Akron Corporation, which had gone bankrupt. In 1974, Eagle Lock lost a bid to the Lori Corporation for a large order for cylinders for the U.S. Postal Service. A few months later the company folded. The Lori Corporation bought the Eagle Lock equipment at a public auction, and the Eagle Lock plants were burned, ending 122 years of lock making.

The Growth of Private Security in America

In 1850, after becoming Chicago's first police detective, Allan Pinkerton, a cooper, started a detective agency. Police departments were limited by jurisdictions. But the private security companies could cross state lines to apprehend fleeing criminals. Pinkerton and other private security companies became famous for catching criminals who crossed jurisdictions. Today, Pinkerton is a subsidiary of Securitas, based in Stockholm, Sweden.

In 1852, William Fargo started Wells, Fargo & Company to ensure the safe transportation of valuables. Today, Wells Fargo is a division of Burns International Services Corporation, which is a subsidiary of Securitas.

William Burns was another security entrepreneur. First, he was a Secret Service agent who directed the Bureau of Investigation, which preceded the FBI. In 1909, he opened the William J. Burns Detective Agency, which became an arm of the American Bankers Association. Today, Burns International Services Corporation is a subsidiary of Securitas.

In 1859, Washington P. Brink also took advantage of the need for safe transportation of valuable freight, package, and payroll delivery. As cargo became more valuable through the years, his service required increased protection. In 1917, following the killing of two Brink's guards during a robbery, the armored truck was initiated. Today, Brink's, Inc., a subsidiary of the Pittston Company, is the world's largest provider of secure transportation services. It also does a lot of business monitoring home alarm systems.

Another major figure in the history of private security in the United States is Edwin Holmes. He pioneered the electronic alarm business. During 1858, Holmes had a hard time convincing people that an alarm would sound on the second floor of a home when a door or window was opened on the first floor. He carried a model of his electronic alarm system door to door. He installed the first of his alarm systems in Boston on February 21, 1858. Sales of his system soon soared, and the first central monitoring station was formed. Holmes Protection Group, Inc., was bought by ADT Security Services, Inc.

Since 1874, ADT Security Services, Inc., has been a leader in electronic alarm services. The company was known originally as American District Telegraph. Today ADT has acquired many security companies. It is a unit of Tyco Fire and Security Services and is the largest provider of electronic security services— serving nearly 3 million commercial, federal, and residential customers throughout North America and the United Kingdom.

Another leader in the private security industry is the Wackenhut Corporation. It provides correctional and human resources services. The company was founded in 1954 by George Wackenhut, a former FBI agent, and today has operations throughout the United States and in more than 50 countries.

Railroads and Labor Unions

The history of private security in the United States is largely related to the growth of railroads and labor unions. Railroads were important for providing the East-West link that allowed the settling of the American Frontier, but the powerful businesses used their domination of transportation to control several businesses, such as coal and kerosene. Farmers had to pay high fees to transport their products by train. The monopolistic practices created a lot of hostility. Citizens applauded when Jessie James and other criminals robbed trains. Because of jurisdictional boundaries, railroads couldn't rely on public police protection. Many states passed laws allowing railroads to have their own private security forces with full arrest powers and the authority to arrest criminals who cross jurisdictions. By 1914, there were 14,000 railroad police. During World War I, railroad police were deputized by the federal government to ensure protection of this vital means of transportation.

The growth of labor unions during the nineteenth century resulted in an increased need for strikebreakers for large businesses. It was a costly business. A bloody confrontation between Pinkerton men and workers at the Carnegie steel plant in Homestead, Pennsylvania, resulted in eight deaths (three security men and five workers). Pinkerton's security forces withdrew. Then the plant was occupied by federal troops.

As a result, the Homestead disaster and "anti-Pinkertonism" laws were passed to restrict private security. Local and state police forces then emerged to deal with strikers. Later, the Ford Motor Company and other businesses were involved in bloody confrontations. Henry Ford had a private security force of about 3500 people, who were helped by various community groups such as the Knights of Dearborn and the Legionnaires. Media coverage of the confrontations gave a negative impression of private security. Before World War II, the Roosevelt administration, labor unions, and the American Civil Liberties Union (ACLU) forced corporate management to shift its philosophy to a softer approach.

World Wars I and II

World Wars I and II brought about an increased need for protection in the United States. Sabotoge and espionage were big threats. Major industries and transportation systems needed, expanded and improved security. The social and political climate in the early twentieth century reflected urban problems, labor unrest, and worldwide nationalism. World War I compounded these concerns and people's fears. A combination of the war, Prohibition, labor unrest, and the Great Depression overtaxed public police forces. Private security helped to fill the void.

By the late 1930s, Europe was at war again, and the Japanese were expanding in the Far East. A surprise Japanese bombing of the Pacific fleet at Pearl Harbor in 1941 pushed the United States into the war, and security again became a major concern for people. Protection of vital industries became critical, causing the United States to bring plant security personnel into the army as auxiliaries to military police. By the end of the war, over 200,000 of these security workers had been sworn in.

Twenty-First Century Security

The 1990s brought the first bombing of the World Trade Center and the bombing of the Murrah Federal Building in Oklahoma City, war with Iraq, crimes over the Internet, increased value of proprietary information, and more violence in the workplace, all leading to an increased need for private security. Current concerns over security notwithstanding, security increasingly has been an issue throughout American history. From the labor struggles of the early twentieth century through the necessary security of World War II and the Cold War to the increased crime of the sixties and seventies, Americans have sought ways to protect their property and stay safe.

In fact, one significant innovation in high-security mechanical locks came in 1967 with the introduction of the Medeco high-security cylinder. From the outside, a Medeco cylinder looks like most other pin-tumbler cylinders, but it works very differently. Figure 1.8 shows how a Medeco cylinder looks inside. The cylinder, made by Roy C. Spain and his team, used chisel-pointed rotating pins and restricted angularly bitted keys that made picking and impressing harder. To open the lock, a key had to not only simultaneously lift each pin to the proper height but also rotate each one to the proper position to allow a sidebar to retract.

The name Medeco was based on the first two letters of each word of the name Mechanical Development Company. The Medeco Security Lock was the largest and most talked about high-security lock. In the early 1970s, the company offered a reward for anyone who could pick open one, two, or three of its cylinders within a set amount of time. In 1972, Bob McDermott, a New York City police detective, picked one open in time and collected the reward. That feat didn't slow the demand for Medeco locks. Much of the general public never heard about the contest and still considered Medeco locks to be invincible. In 1986, Medeco won a patent infringement lawsuit against a locksmith who was making copies of Medeco keys. That ruling stopped most other locksmiths from making the keys without signing up with Medeco. The patent for the original

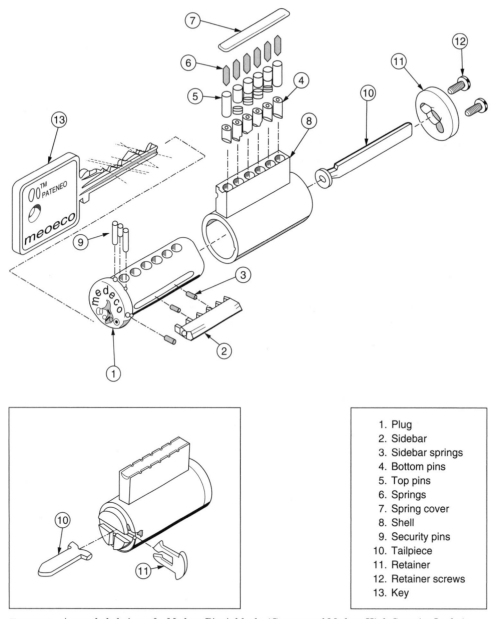

Figure 1.8 An exploded view of a Medeco Biaxial lock. (*Courtesy of Medeco High Security Locks.*)

1. Plug
2. Sidebar
3. Sidebar springs
4. Bottom pins
5. Top pins
6. Springs
7. Spring cover
8. Shell
9. Security pins
10. Tailpiece
11. Retainer
12. Retainer screws
13. Key

Medeco key blank ran out, and now anyone can make keys for those cylinders. In 1988, the company received a new patent for its Biaxial key blanks. The big difference is that the Biaxial brand gave Medeco a new patent (which can be helpful for preventing unauthorized key duplication).

The company's latest locks use its patented Medeco 3 technology. The most obvious improvement in Medeco 3 is the addition of a third locking element within the cylinder and on the key. With three distinct locking points, Medeco 3 continues to resist picking while increasing to 13 billion the number of unique codes available for master-key systems.

Brink's Home Security was founded in 1983 as an affiliate of Brink's, Inc.— the world's largest provider of secure transportation services. Its headquarters are in Irving, Texas, where the National Service Center (NSC) provides alarm monitoring and other customer services. Today, Brink's Home Security monitors home security systems for more than 1 million customers in over 200 markets in the United States and Canada.

Early in the twenty-first century, on September 11, 2001, terrorists hijacked airplanes and attacked the World Trade Center and the Pentagon, which resulted in about 3000 deaths. Such a bold surprise attack shows how challenging security is these days. It requires a new way of thinking about security. Security professionals need to consider how to prevent not only a variety of accidents, disasters, crimes, and fire but also terrorism and bioterrorism. These days it's critical for loss-prevention officers to have improved education and training so that they can face current challenges professionally and with creativity and imagination. This book will help you to better face the myriad loss-prevention challenges.

2

Strengthening Doors

Whenever I'm asked about home security, the questioner seems to be concerned mainly with door locks. Rarely does anyone think to ask about making the doors themselves stronger. I've been in homes where people were asking me to recommend a high-security lock for a door that looked as though it would fall down if anyone knocked on it too hard.

It's important to understand that a lock is just a device that fastens a door to the door frame. Using a good lock on a thin-paneled door or on a door with weak hinges is like using a heavy-duty padlock to secure a paper chain. Before worrying about a good lock, be sure you have a strong door and frame.

If burglars can't get in through your doors, where do you think they might try next? To keep burglars out, you need to secure all your points of entry—windows, skylights, sky roofs, and other wall and roof openings, as well as the doors. In this chapter I'll review how burglars might get through your home or office doors, how to make them more secure, and how to choose and install new doors.

How Burglars Can Get Through Doors

Burglars are not invited guests. Don't allow them to enter your home as though they were. Be aware of burglars' secrets for getting past doors and what you can do to keep them out.

Removing the hinges

If a door's hinges can be seen from the exterior side, a burglar may be able to remove the hinges and open the door without touching the lock. Most door hinges consist of two metal leaves (or *plates*)—each with "knuckles" on one edge—and a pin that fits vertically through the knuckles when they're aligned and holds the leaves together.

The hinge pins often can be pulled out with little difficulty, and the door then becomes disconnected from the door frame. Burglars who remove a door in this way can place the door back on its hinges on the way out, and you may never know how they got in (and your insurance company may not want to pay your claim). One way to prevent this type of entry is to use hinges with nonremovable pins—pins that are either welded in place or secured by a setscrew or retaining pin.

If you don't want to replace your door hinges, you can install hinge enforcers. These small metal devices attach to the hinge and the door frame to block the door's removal even when the hinge pins have been removed. A package of hinge enforcers costs about $5.

Prying off stop molding

If your door's hinges can't be seen from outside, your next concern should be your door's stop molding. Stop moldings are the protruding strips (usually about $1/2$ inch thick) that are installed on three sides of a door frame—the lock side, the hinge side, and the header (top). They stop the door from swinging too far when you're closing it.

Depending on which way the door swings, a person standing outside the door will be able to see either the hinges or the stop molding. Some stop moldings are simply thin wooden strips tacked to the frame and can be pried off easily. By removing the lock-side strip, a burglar exposes the bolt and thus makes it easier to attack the lock. To solve the problem, you can remove the stop moldings and reinstall them using wood glue and nails so that they can't be pried off easily. When you buy or make a new door and frame, be sure that its stop molding is milled as an integral part of the jamb.

Kicking doors down

A common way burglars get through doors is by kicking them in. If either the strike plate on the door jamb or the lock edge of the door is weak, a strong kick will break the door open. Short of getting a new door, the best way to solve a weak-door problem is to install door reinforcers (Figures 2.1 and 2.2). They usually cost less than $20 each.

One type of door reinforcer is a U-shaped metal unit designed to wrap around the door edge near the lock. Designs are available for doors with one or two locks. To install this type of reinforcer, first remove the locks from your door. Position the unit so that the lock holes are fully exposed, and screw it firmly into place. Then install the locks.

Weak door frames also can be strengthened. A popular reinforcer for door frames is the high-security strike box, a heavy-gauge steel box (also called a *box strike*) with 3-inch-long screws or rods that protrude through the door jamb and into a wall stud (Figure 2.3). The strike box is stronger than the more commonly used thin, flat strike plates that are fastened only to the jamb using small wood screws.

Figure 2.1   A standard door reinforcer. (*Courtesy of M.A.G. Manufacturing.*)

Impersonating other people

No matter how much hardware you have on your doors, it won't keep out burglars who trick you into letting them in. Burglars often pretend to be delivery persons, police officers, or meter readers, and trusting people immediately open the door for them. Always ask to see identification, or ask that the name and

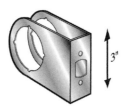

Figure 2.2   A designer door reinforcer. (*Courtesy of M.A.G. Manufacturing.*)

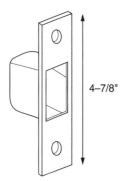

4–7/8"

Figure 2.3   A strike box fits into the door frame. (*Courtesy of M.A.G. Manufacturing.*)

full address on the delivery be read to you before you open the door. If your door has no glass pane, install a wide-angle door viewer that lets you see anyone standing outside (Figure 2.4). Better models allow you to see in several directions. Door viewers range in price from under $5 to over $50.

Don't depend on a door chain to let you see who's at your door. Door chains provide a false sense of security. A person who intends to break in can easily break off the chain by pushing sharply against the door.

Prying open sliding-glass doors

A sliding-glass door (sometimes called a *patio door*) usually consists of two glass panels (or *sashes*) that slide along tracks. Doors of this type are especially vulnerable because their frames and locks are weak. A sliding-glass door can be forced open by prying the sliding panel away from the door frame. You can thwart such an entry technique by inserting a length of wood into the metal track on which the door slides. The wood acts as an obstruction; the door cannot be slid open from the outside. You can easily reopen it from the inside by removing the wood. Various styles of sliding-door barriers are sold in hardware stores and home-improvement centers. Some models are designed to let you keep the door open a few inches for ventilation. Another way to prevent a burglar from prying the sliding door from the door frame is to use a lock that holds the frames together (Figure 2.5).

Figure 2.4   A door viewer lets you see who is at your door without opening the door. (*Courtesy of M.A.G. Manufacturing.*)

Figure 2.5 By locking the panels together, you can prevent a burglar from prying the panels apart. (*Courtesy of M.A.G. Manufacturing.*)

Burglars can also defeat sliding-glass doors by using a pry bar to lift the sliding sash out of its lower track. You can install screws or antilift plates at the top of the door to resist this entry technique. These devices create an obstruction between the door and frame. You can buy a package of antilift plates for less than $5.

To install an antilift plate, first close and lock the door. Position the plate so that it is on the center of the frame and butts against the upper track; then screw the antilift plate into place. (You may need to use an awl to puncture screw holes into the frame.) An antilift plate isn't adjustable for clearance all along the track. Antilift screws also resist attempts to lift a sliding door out of its lower track, but they are adjustable to allow for door clearance all along the track. The screws cost about the same as the plates and need to be installed about every 12 inches along the length of the center of the upper track.

Breaking in through garage doors

Don't ignore your garage doors. Burglars know that a typical garage contains cars, bikes, lawn mowers, tools, and other easy-to-sell items. A garage that's attached to a home usually provides easy access to the home. The most secure main garage doors are made of steel, require an automatic door opener, and have no glass panels.

Panels of any material weaken a door, but glass panels are an especially poor feature in garage doors. They can be broken easily, and they allow a burglar to see what's in the garage. If you have glass in a garage door, you might want to paint over the glass from the inside.

To reinforce panels of any material on a wood door, you can install an angle iron. Even if a burglar breaks a panel, the angle iron will block entry.

In addition to securing the main door of a garage, you should reinforce any door that allows passage from the garage to your home. That "inside" door should be as secure as any exterior door. Burglars who are able to drive into your garage and enter your home through the garage entrance will be unseen while they load your possessions into their car. On another topic, check your local building code. It may require that a door connecting a home to a garage be fire resistant.

Choosing a New Door

In most homes, the style of door depends on the structure's architecture. You'll probably want doors that complement your home's design, but regardless of the style, you'll need to decide on the materials of the door unit.

Typically, an exterior door is connected to its frame by metal hinges on one side and a lock bolt on the other. The frame consists of various sections: a head jamb (along the top), two side jambs, stop moldings on the top and sides, and a sill or threshold (along the bottom). Although the door and frame don't have to be made of the same material, they usually are. Commonly used materials include steel, wood, aluminum, fiberglass, polyvinyl chloride (PVC) plastic, and glass.

Steel doors offer the best protection against fire and break-in attempts. They also offer superior insulation, which helps to keep energy costs down. Many steel doors are beautifully designed to look like expensive wood doors. Kalamen doors consist of metal wrapped around a wood core; they provide good security when installed with a strong frame.

Fiberglass, a strong material that can be made to look like natural wood, offers good resistance to warping and weathering; it's especially useful near pools and saunas or in damp areas. Some types of fiberglass can be stained and finished. Like fiberglass, PVC plastic is strong and isn't affected much by water. However, the plastic surface can be hard to paint.

Aluminum and glass are used together, mostly for sliding-glass doors. Although glass can make a door look nice and admits light to the interior, it also makes the door less secure.

Among wood doors, the solid-core hardwood types are best. They consist of hardwood blocks laminated together and covered with veneer. A hollow-core door provides minimal protection; it consists of two thin panels over cardboard-like honeycomb material. You can recognize a hollow-core door by knocking on it; it sounds hollow. If a burglar kicked a hollow-core door, his or her foot would go through it.

There's an easy way to reinforce a hollow-core door if aesthetics aren't important. You can clad the exterior side with 12-gauge (or thicker) sheet metal attached with 5/16-inch-diameter carriage bolts. The bolts should be placed along the entire perimeter of the door about 1 inch in from the door's four edges. Space the bolts about 6 inches apart, and secure them with nuts on the interior side of the door. If after installing the metal you find that the door is too heavy to open and close properly, you may need to remove the hinges and install larger ones.

Another important factor affecting door strength is whether it's flush or paneled. A flush door is flat on both sides and is plain-looking. A paneled door has surfaces of varying thicknesses and can be very attractive. The panels may be metal, wood, glass, or a combination of materials. Because the panels usually are thinner and weaker than the rest of the door, they make the door more vulnerable to attack.

You can buy a door, side jambs, trim, threshold, and sill as separate parts, or you can get all the parts together in a single package—a door kit with precut jambs and sills, which is easier to install. You also can buy a prehung (or pre-assembled) door, ready to be fastened to the rough opening.

Many modern door units come with sidelight panels (small vertical windows along the sides). If the sidelights might allow a burglar to climb through or to reach in for the lock, they should be made of or lined with a break-resistant material such as plastic.

M.A.G. Manufacturing History and Products

M.A.G. Manufacturing is a leading maker of door strengthening. One evening in the late 1960s, Howard Allenbaugh, founder of M.A.G. Manufacturing, was listening to a custodian gripe about a recent rash of student break-ins at a local college. As the custodian complained about not knowing how to prevent students from kicking in locked doors, Howard sketched out a drawing of the now famous M.A.G. door reinforcer, a U-shaped metal plate that is placed around the door knob and lock to strengthen the door at its weakest point, thus resisting kick-ins. More than 30 years since its inception, the door reinforcer is now a popular security feature in hotels, restaurants, apartments, offices, and homes.

Since the design of the door reinforcer, M.A.G. Manufacturing has grown from one employee to 60 employees in the United States and now offers over 1000 security products. Located in southern California, M.A.G. is dedicated to providing consumers with innovative security and safety products, as well as fulfilling social and ethical responsibilities. The company is involved in many humanitarian efforts, including the national Safe At Home campaign, which provides financial support to assist child-safety advocacy groups in their efforts to educate parents on at-home safety practices.

Since the death of Howard Allenbaugh in 2003, M.A.G. Manufacturing has been lead by the founder's son, Mark H. Allenbaugh—a practicing attorney and former business and professional ethics professor at George Washington University.

M.A.G.'s business was built by locksmiths seeking solutions to their workplace challenges. This same core relationship is the foundation of M.A.G.'s innovation and design strategy to be the locksmith's security and hardware provider. Providing commercial-grade security hardware remains at the forefront of M.A.G.'s focus, as it has been since its inception.

M.A.G Manufacturing regularly exhibits at several security conventions, including those of the Associated Locksmiths of America, the Door and Hardware Institute, and the Security Hardware Manufactures Association. The company's most popular security products among locksmiths are door reinforcers, door edge guards, latch guards (Figure 2.6), key lockouts, strikes, hole covers (Figure 2.7), and filler plates (Figure 2.8). The door reinforcer comes in several finishes:

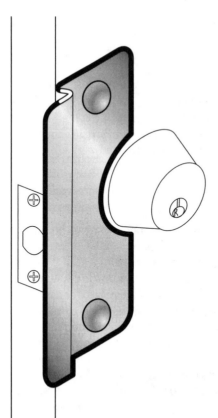

Figure 2.6 A latch guard protects the lock's latch. (*Courtesy of M.A.G. Manufacturing.*)

polished brass, antique brass, satin bronze, oil-rubbed bronze, satin nickel, aluminum, polished stainless steel, and stainless steel.

M.A.G. offers residential and commercial door reinforcers. The residential door reinforcers come in several finishes: brass, chrome, and durandoc. The commercial models come in brass, chrome, duranodic, prime coat, aluminum, and stainless steel. The primary difference between the two types is the latch preparation. Commercial door reinforcers have a recessed-edge preparation to accept the lock latch. This gives the professional installer the flexibility of interchanging locks or repairing the lock without having to remove the door reinforcer to access the latch. Residential door reinforcers have a flat-face design that makes installation of the plate much easier by installing over the latch. However, this isn't intended for applications where access to the latch is required or probable. M.A.G. also makes several models of door reinforcers exclusively for commercial use on electronic locks.

M.A.G.'s Uni-Force door edge guard provides the same strength of the door reinforcer, but the Uni-Force is designed to be a more aesthetically pleasing preventative product than the door reinforcer by adding an aesthetic element and an element that also covers damage caused by kick-ins or marks left by previous lock's preparation.

Figure 2.7 As the name implies, a hole cover is used to cover lock holes. (*Courtesy of M.A.G. Manufacturing.*)

M.A.G.'s latch guards prevent someone from spreading the door and frame. There are models for in-opening doors and some for out-opening doors. Residential latch guards come in three finishes: brass, chrome, and duranodic. Commercial models come in brass, chrome, duranodic, prime coat, and aluminum.

Figure 2.8 A filler plate. (*Courtesy of M.A.G. Manufacturing.*)

Another popular M.A.G. security product is the Key Lock-Out. It is a temporary lockout solution that uses the lockset keyway as the deterrent. Picture a key that has been broken off in the keyway, obstructing any other keys from entering. With the Key Lock-Out, there is a patented tool that allows this "broken" key bow to be removed when the need for the lockout has passed or a professional installer changes the keyway. To engage the Key Lock-Out, simply insert the "key" into the lock, and the detachable tip will remain inside the keyway undetected. Once installed in a lock, it completely prevents a working key from operating the lock. To remove the detachable tip, insert the key again, and the tip will reconnect and slide out, making the keyway operable.

"Key Lock-Out is an essential tool for many nerve-wracking situations, such as losing one's house keys," says Mark Allenbaugh. "Key Lock-Out is durable and can be used until lost keys are found or until a locksmith can change the lock." Key Lock-Out is also an effective tool for apartment and building managers, construction-site guards, business owners, and nursing homes and hospitals.

M.A.G. also offers several popular strikes. Its Adjust-a-Strike (Figure 2.9) is a T-strike that uses tongue-and-groove plates to give it its patented adjustable feature. This product has a great demand because of not damaging doors, frames, and entire buildings for that matter. The T-strike lets you extend the strike past decorative moldings or masonry structures, allowing the latch to pass by without damaging the molding or masonry. M.A.G.'s Double-Strike is designed to equally distribute strength between the deadbolt and knob latches.

M.A.G's newest security products are application-specific welded stud guards. The studs are concealed on the outside, making the latch guard more aesthetically pleasing. They come in various models specific to electronic locks, levers, mortise locks, electronic strikes, and key-in-knob locks.

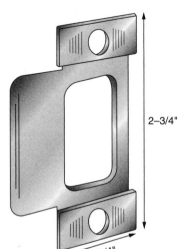

2–3/4"

1–1/4"

Figure 2.9 An adjustable strike. (*Courtesy   of   M.A.G. Manufacturing.*)

Certificate of Completion

This certificate acknowledges that

Jennifer Weaver

has successfully completed the M.A.G.-Certified Locksmith Distributor course.

November 16, 2005
Date Issued

Mark Allenbaugh
President

Figure 2.10   If you pass M.A.G.'s online quiz, you will receive a certificate. (*Courtesy of M.A.G. Manufacturing.*)

In 2006, M.A.G. introduced its M.A.G. PRO line of commercial-grade security hardware products developed specifically for locksmiths. M.A.G. doesn't sell to end-users. Its products are only sold through locksmiths and other distributors, such as Ace Hardware, Home Depot, True-Value Hardware, and Do It Best retail outlets.

M.A.G. Manufacturing offers vendor certification to locksmiths who demonstrate exceptional knowledge of the company's product lines. The certification process involves registering your company online and passing a short online quiz. If you pass the test, you'll receive a certificate suitable for framing (Figure 2.10). To become certified or to learn more about M.A.G. products, go online to www.magmanufacturing.com.

3

Exit Door Devices

To comply with building and fire codes, businesses and institutions often have to keep certain doors as emergency exits that can be opened easily by anyone at any time. (This is to help prevent not having enough quick ways out during a fire or other emergency.) In some cases, however, the doors that must remain easy to exit also need to be secured from unauthorized use (such as when the door may allow shoplifters to slip out unnoticed).

Most institutions and commercial establishments use emergency exit door devices as a cost-effective way to handle both matters (Figures 3.1 and 3.2). Such devices are easy to install and offer excellent money-making opportunities for locksmiths.

Typically such devices are installed horizontally about 3 feet from the floor and have a bolt that extends into the door frame to keep the door closed. They also usually incorporate either a push bar or clapper arm that retracts the bolt when pushed.

Some models provide outside key and pull access when an outside cylinder and door pull are installed. In these models, entry remains unrestricted from both sides of the door until the deadbolt is relocked by key from inside or outside the door.

Many emergency exit door devices feature an alarm that sounds when the door is opened without a key. The better alarms are dual piezo (double sound). Other useful features to consider on an emergency exit door device include a choice of the hand on which it is installed (nonhanded models are the most versatile), the length of the deadbolt (a 1-inch throw is the minimum desirable length), and special security features (such as a hardened insert in the deadbolt).

This chapter explains how to install, operate, and service some of the models manufactured by Alarm Lock Systems, Inc. Much of the information also applies to most other popular models. This chapter also tells how exit devices are tested and rated.

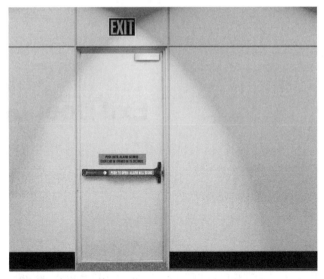

Figure 3.1   An emergency exit device. (*Courtesy of Detex Corporation.*)

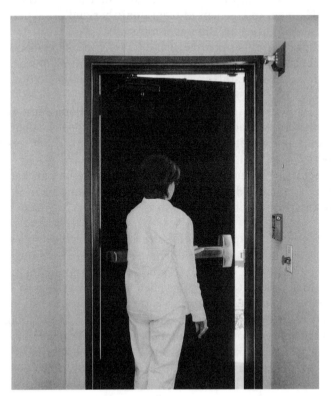

Figure 3.2   An emergency exit device makes a locked door easy to open in an emergency. (*Courtesy of Detex Corporation.*)

Distinctive Exit Devices Also Provide Safety*

Exit devices now come in such a wide range of styles (such as push bar and cross-bar) and finishes (such as aluminum, stainless steel, brass, and bronze) that they complement doors in a variety of applications and can enhance the design integrity of a building, whether it is traditional or modern. More important, however, is an exit device's safety function.

You may use the same exit devices day after day without realizing that some are panic devices and others are fire-rated devices. The four common types of both panic and fire-rated devices are mortise, rim, surface vertical rod, and concealed vertical rod. They have the same designs and finishes and, for the most part, function similarly. You can get both panic and fire-rated devices for light-, medium-, or heavy-duty traffic installations.

Even though you may not be able to tell a panic device from a fire-rated device at a quick glance, there are big differences in application, material, and product testing.

Which Exit Device to Choose

Panic devices, which provide no guaranteed protection against fire, are for life safety; they allow a person to get out through the door quickly. Conversely, fire-rated exit devices are designed to keep fire from getting in through the door. A fire-rated device is designed to stay latched so that the door doesn't open during a fire.

There is a quick way to tell the difference between panic and fire-rated exit devices. Panic (life safety) devices usually have dogging. In a school, for instance, a custodian can use a dogging key to lock the latchbolt in. Since the latchbolt cannot project when it's dogged, the door basically becomes a push-pull door, allowing people easy entrance and exit. Fire-rated exit devices are not allowed to have dogging. Every time the door shuts, it latches, creating a secure fire wall. Such exit devices have a label somewhere in view that says that they are fire rated.

The type of door device installed depends on specifications set by local building codes, the fire marshall, or whomever has jurisdiction. Most often, internal exit devices are fire-rated, and external doors have panic devices. External doors have panic exit devices to allow quick exit, whereas internal doors have fire-rated exit devices to create a safe barrier against any fire outside the door.

The most essential difference between fire-rated and panic exit devices is the material used to make the latchbolts. Latchbolts in fire-rated hardware generally are made of stainless steel because of the material's ability to withstand fire. On the other hand, panic devices are made of forged brass, which can't stand up to fire as well as stainless steel but is more economical and meets the requirements of panic devices. You can use the same trim on both fire-rated and panic exit devices.

---

*This section was written by Ashley Rolfe.

Many manufacturers perform in-house testing of panic devices and use Underwriters Laboratories (UL) to test fire-rated devices. The American National Standards Institute (ANSI) and the Builders Hardware Manufacturers Association (BHMA) set the standards, and UL enforces them. UL doesn't rate quality but simply ensures that manufacturers meet industry guidelines.

Currently, ANSI and BHMA don't set stringent guidelines for manufacturers to follow when setting up their panic device testing facilities, so each apparatus is a little different than others. Because most manufacturers have their own testing apparatus for panic devices, a UL field representative comes to the factory to supervise the testing of each type of product. A randomly selected exit device is attached to a door that is specially mounted to I-beams or similar rigging for testing. UL then puts the device through a series of tests.

Exit tests (grades 1 and 2)

With the door latched, a maximum of 15 pounds force is applied against the bar, and the door swings open. The device also must work when the force is applied anywhere along the bar in the direction of the swing. The door also must swing open when a maximum 250 pounds force is applied against the door and a maximum of 50 pounds force is applied against the bar.

Security test (grade 1)

A pull of 400 pounds force is applied to the door. The door shouldn't open, and the device should function properly after the test.

Inside pull test (grade 1)

Here, 400 pounds force is applied at the center of the locked bar in the opposite direction of the swing. After the force is removed and the bar is released, the device should still function, and no damage should be apparent.

Push test (grade 1)

With the touchbar free and the door fastened securely to prevent it from swinging open, 400 pounds force is applied at the center of the bar in the direction of the swing. The device should work when the test is finished, and there should be no apparent damage.

Torque to retract latchbolt test

A torque load is applied to the knob, lever, or turn to retract the latchbolt of an unlocked device. The torque is applied slowly until the latchbolt clears the strike. For thumbpieces, the load is applied at a point $1/4$ inch, from the end of the thumbpiece until the latchbolt clears the strike. Where applicable, the test is repeated in the opposite direction Values can't exceed 15 pounds force per inch for levers, 15 pounds force per inch for thumbpieces, and 22.5 pounds force per inch for turns.

Force to latch door test

Force is applied to the face of the door 1 inch from the lock edge of the door and on the centerline of the latchbolt when the door is just clear of the latch contacting the lip of the strike. The door is closed slowly by pushing the force gauge against the door until the latchbolt engages the strike fully. The maximum measured force to fully latch the door cannot exceed 4.5 pounds force.

Trim durability test

Outside grade 1 trim is tested for at least 250,000 cycles and outside grade 2 trim for at least 100,000 cycles. The torque to retract latchbolt test is then repeated. The opening values should not exceed by more than 20 percent those in the earlier test.

Cylinder test

The cylinder used on an exit device must meet ANSI/BHMA requirements for auxiliary locks and associated products. The cylinder is to have been tested in other locks and have met the applicable requirements. Additionally, the torque necessary to retract the latchbolt(s) by key or to lock and unlock the trim by key must be measured.

Unlocked outside knob or lever torque test

With the device in the latched position, 250 pounds force per inch torque to grade 1 knobs or turns, 150 pounds force per inch torque to grade 2 knobs or turns, 450 pounds force per inch torque to grade 1 levers, and 225 pounds force per inch torque to grade 2 levers is applied. At the end of the test, the maximum torque to retract the latchbolt cannot exceed 18 pounds force per inch for knobs, 27 pounds force per inch for turns, or 50 pounds force per inch for levers. Devices should operate in all respects.

Locked outside knob torque test

A torque is applied to a locked knob or thumbturn of 300 pounds force per inch for grade 1 and 150 pounds force per inch for grade 2 knobs. After the load is released, the torque to retract the latch cannot exceed an 18 pounds force per inch torque for knobs or 27 pounds force per inch torque for turns.

Locked outside thumbpiece test

A load of 150 pounds force for grade 1 and 150 pounds force for grade 2 is applied at a point 1 inch from the end of the thumbpiece until the latchbolt clears the strike. After the load is released, the force to retract the latch cannot exceed a 50 pounds force per inch torque.

Axial-load test

A load dynamometer force of 400 pounds force for grade 1 and 300 pounds force for grade 2 is applied to the outside knob, lever, turn, or thumbpiece grip along the axis of the trim perpendicular to the face of the door to load the latchbolt against the strike. After the load is released, the force to retract the latch can't exceed 18 pounds force per inch for knobs, 27 pounds force per inch for turns, or 50 pounds force per inch for levers. The load is applied at the highest points of knobs and for levers 1 inch from the face of the door or escutcheon, if present.

Vertical-load test

A load dynamometer force of 360 pounds force for grade 1 and 250 pounds force for grade 2 is applied vertically downward to the outside knob, lever, or turn and perpendicular to the trim axis. After the load is released, the force to retract the latch can't exceed 18 pounds force per inch for knobs, 27 pounds force per inch for turns, or 50 pounds force per inch for levers. The load is applied at the highest point of knobs and for levers 1 inch from the face of the door or escutcheon, if present.

Outside knob or lever crush test

A knob or lever is positioned in a tensile loading device and compressed with 1000 pounds force. Deformation cannot exceed 10 percent for grade 1 or 25 percent for grade 2 knobs or levers.

Outside rose and escutcheon dent test

An 8-ounce projectile is dropped from a height of 12 inches in a tube between the edges of the rose or escutcheon. The depth of the dent cannot exceed 0.075 inch for grade 1 or 0.100 inch for grade 2 roses or escutcheons.

Outside rose and escutcheon deformation test

Outside trim is mounted to simulate its installation on a door. A compression of 650 pounds force for grade 1 or 560 pounds force for grade 2 is applied on the horizontal centerline of the rose or escutcheon assembly. The sidebars to apply the compression load must be 6 inches long. When testing escutcheons, the vertical centerline of the 6-inch-long bars must be opposite the centerline of the escutcheon. Deformation should not exceed 10 percent. Finish tests ensure consistent trim quality. Trim parts must meet minimum values given.

Salt-spray test

Clear-coated materials are exposed for 96 hours, lock fronts and strikes for 24 hours, uncoated base materials for 200 hours, and painted and powder-coated materials for 96 hours.

Humidity test

Clear-coated materials are exposed for 240 hours, lock fronts and strikes for 48 hours, and painted and powder-coated materials for 240 hours.

Pencil hardness test

Materials with organic coatings are tested in lieu of the Tabor abrasion test at the manufacturer's option.

Ultraviolet (UV) light and condensation test

Exterior-grade finishes are tested for 8 hours of UV light at 60°C and 4 hours of humidity at 50°C for a total of 144 hours.

Tabor abrasion test

This test is conducted in lieu of the pencil hardness test at the manufacturer's option for 500 cycles.

After all these tests and others, the UL representative takes apart the exit device and checks for wear points. The most obvious spots for wear are the strike and latchbolt. Manufacturers commonly go through the same procedure on both metal and wooden doors. UL also will check the follow-up service books on each product to ensure that the descriptions and drawings in the book match the tested device.

The examiner then checks to ensure that the device the manufacturer is having tested is the same product the manufacturer is selling. This often means a trip to the factory warehouse to check inventory against the test product. The inspector won't necessarily run a randomly selected product through a cycle test but may conduct a parts inspection, again to ensure that the parts match the drawing. UL also occasionally will take a panic device off the shelf to test it.

Most likely, if the testing does not meet ANSI standards, the manufacturer makes a change that UL finds acceptable. No American manufacturer of exit devices will sell their products without a UL label.

How a Fire Rating Is Determined

To have their hardware fire rated, manufacturers go to the UL fire division in Chicago, where the device is attached to a door. The door is hung on a brick fire wall so that the opposite side of the exit device faces a powerful furnace.

The hardware is subjected to the intense fire from the furnace for a period of time depending on the rating the manufacturer seeks. There isn't much left of the door when the test is finished, and the exit device isn't in very good shape either. However, the latchbolt still should maintain engagement with the strike.

If the door doesn't stay locked throughout the test, the device fails. In that case, the manufacturer's engineers either go back to the drawing board or make relatively minor adjustments and retest the device. The manufacturer's cost for each fire test runs to tens of thousands of dollars.

Fire-rated doors have several different grades and tests. A-label devices are fire rated for 3 hours, which means that they withstood 3 hours of a fire while staying locked. B-label devices have a $1\frac{1}{2}$-hour fire rating, and C-label doors have a 45-minute rating. Some buildings may need only C- and B-label devices, whereas others exclusively require A-label hardware.

Most manufacturers would test double egress doors as standard, therefore complying with the test requirement of burning the active and inactive door to the fire. Those that choose not to do so would be subject to two tests—one with the device to the fire and the other with the device away from the fire.

Although panic devices must pass a 250,000-cycle test minimum, fire-rated devices, which often are subjected to less routine wear and tear than panic hardware, must only pass a 100,000-cycle test. Many manufacturers often try to do two, three, or even four times the minimum when doing their own testing.

Every manufacturer is on a testing schedule. For instance, a manufacturer may have five different series of exit devices, and each must be tested every 2 years. No matter how long a product has been on the market, it must be retested to ensure that it still meets ANSI standards.

No industry standard exists for the apparatus used in testing panic devices. Each manufacturer does it his or her own way. There is, however, a consensus to reach the same conclusions.

The external appearance of exit devices hasn't changed much in several years and may not for years to come. Thus the average person probably never will notice the difference between panic and fire-rated exit devices. But the differences are real and essential. And engineers look continually for ways to improve quality and endurance.

ANSI and BHMA also continue to look for ways to strengthen industry standards. Even though the industry is essentially self-governing, the stringency and frequency of testing ensure the performance and safety of exit devices made in America.

Pilfergard Model PG-10

One of the most popular emergency exit door devices is the Pilfergard PG-10 (Figure 3.3). It has a dual-piezo alarm, can be armed and disarmed from inside or outside a door, and is easy to install on single or double doors.

Figure 3.3 The Pilfergard PG-10 is a popular emergency exit door device. (*Courtesy of Alarm Lock Systems.*)

The device is surface mounted approximately 4 to 6 feet from the floor on the interior of the door with a magnetic actuator on the frame (or vice versa). It is armed or disarmed with any standard mortise cylinder, which is not supplied. Opening the door, removing the cover, or any attempt to defeat the device with a second magnet, when the device is armed, activates the alarm.

Installation

To install the Pilfergard PG-10,

1. Remove the cover by depressing the test button, and lift the cover out of the slot.

2. Mark and drill holes per the template directions and drill sizes (five for the alarm unit, two for the magnetic actuator).

3. For inside cylinder installation, proceed to the next step. For outside cylinder installation, drill a $1^1/_4$-inch hole as shown on the template. Install a rim-type cylinder through the door, and allow flat tailpiece to extend 1 inch inside door. Position the cylinder so that the keyway is vertical (horizontal if the PG-10 is installed horizontally). Hold the PG-10 in position over the mounting holes, and note that the outside cylinder tailpiece is centered in the clearance hole in the base of the PG-10 (rotate the cylinder 180 degrees if not). Tighten the outside cylinder mounting screws.

4. Install the PG-10 and magnetic actuator with seven screws.

5. Install the threaded $1^1/_4$-inch mortise cylinder in the PG-10 cover using the hardware supplied (Figure 3.4). The keyway must be horizontal so that the tailpiece extends toward the center of the unit when the key is turned.

6. Move the slide switch to off. (Figure 3.5) Connect the battery. Hook the cover on the end slot, and secure it with the two cover screws. *Note*: One of these screws acts as the tamper alarm trigger, so be sure that the screws are fully seated.

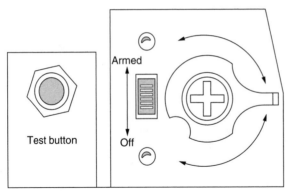

Figure 3.4   Install the mortise cylinder. (*Courtesy of Alarm Lock Systems.*)

Caution: When installing the PG-10 on a steel frame, it might be necessary to install a nonmagnetic shim between the magnetic actuator and the frame. This prevents the steel frame from absorbing the magnet's magnetic field, which could cause a constant alarm condition or occasional false alarms. The shim should be $1/2 \times 2^1/_2 \times 1/_8$ inch thick and may be constructed from plastic, Bakelite, or aluminum.

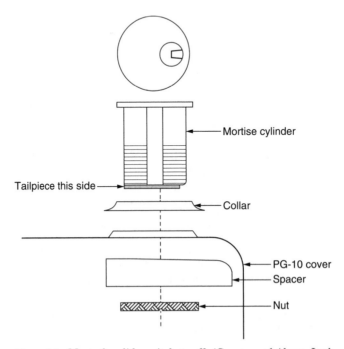

Figure 3.5   Move the slide switch to off. (*Courtesy of Alarm Lock Systems.*)

Testing and operating

To test and operate the PG-10, first depress the test button with the slide switch in the off position. Horns should sound. To test using the magnetic actuator, proceed as follows:

1. Close the door.
2. Arm the PG-10 by turning the key clockwise 170 degrees.
3. Open the door; the alarm should sound.
4. Close the door; the alarm should remain sounding.
5. Silence the alarm by turning the key counterclockwise until it stops.
6. Close the door and rearm the PG-10 by turning the key clockwise until it stops.

The unit should be tested weekly using the test button to ensure that the battery is working. The test button only operates when the PG-10 is turned off.

Pilfergard PG-20

The Pilfergard PG-20 is a sleek, modern version of the PG-10. The PG-20 is designed to fit all doors, including narrow stile doors. It has a flashing light-emitting diode (LED) on its alarm and is as easy to install as the PG-10.

Installation

To install the PG-20, remove the cover from the mounting plate. Install the mortise cylinder (which is not supplied), keeping the key slot pointing down in the six o'clock position. Screw the lock ring on the cylinder.

Select the proper template for the specific type of door. Mark and drill 9/64-inch-diameter holes per the template directions on the door and jamb. This requires four holes for the mounting plate and two holes for the magnetic actuator. *Note*: Certain narrow stile doors require only two holes for the mounting plate. For the outside key control only, drill a $1^1/_4$-inch-diameter hole as shown on the template.

Install a rim-type cylinder (not supplied) through the door, and allow the flat tailpiece to extend $^5/_{16}$ inch beyond the door. Position the cylinder, keeping the key slot pointing down, in the six o'clock position. Tighten the cylinder screws. Knock out the necessary holes from the mounting plate, and install the plate on the door with no. 8 sheet metal screws (supplied). Make sure that the rim-cylinder tailpiece fits in the cross slot of the ferrule if the outside cylinder is used. Install the magnetic actuator on the doorjamb with two no. 8 sheet metal screws (supplied). *Note*: It is sometimes necessary on steel frames to install a nonmagnetic shim between the magnetic actuator and the frame. This prevents the absorption by the steel frame of the magnet's magnetic field, which could cause a constant or occasional false alarm. The shim should be $^1/_2 \times 2^1/_2 \times ^1/_8$-inch-thick nonmagnetic material such as plastic, Bakelite, or rubber.

Make sure that the battery is connected. Cut the white jumper only on the side of the unit on which the magnet will not be installed. The unit will not function otherwise. Select the options desired as follows.

Cut the yellow jumper for a 15-second entry delay. (The alarm will sound 15 seconds after any entry through the door if the unit is armed.) To avoid the alarm on entry, reset the unit within 15 seconds. This feature is used for authorized entrance.

Cut the red jumper for 15-second exit delay. (The unit will be activated after 15 seconds each time the unit is turned on with a key.) This feature allows authorized nonalarmed exit.

Place the shunt jumper plug in position C or 2. If the jumper is in position C, the alarm will sound continuously until the battery is discharged. If the jumper is in position 2, the alarm will be silenced after 2 minutes, and the unit will rearm. However, the LED will continue to flash, indicating an alarm has occurred.

Install the cover on the mounting plate, making sure that the slide switch on the printed-circuit board fits into the cam hole. Secure the cover with the four screws supplied.

Testing

To field test the PG-20, proceed as follows:

1. With the door closed, turn the key counterclockwise until it stops (less than one-quarter turn). The unit is now armed instantaneously or delayed.

2. Push the test button, and the alarm should sound, verifying that the unit is armed.

3. Open the door and then close it. The alarm should sound instantaneously or delayed.

4. Note the pulsating sounder and flashing LED indicating an alarm.

5. To reset the unit, turn the key clockwise until it stops (less than one-quarter turn). The unit is now disarmed. This can be done at any time.

Exitgard Models 35 and 70

To install Exitgard Models 35 or 70 panic locks, proceed as follows:

1. With the door closed, tape the template to the inside face of the door with the centerline approximately 38 inches above the floor. Mark all hole locations with a punch or awl. Note that the keeper is surface installed on the jamb for single doors and on the other leaf for pairs of doors.

2. If an outside cylinder is being used, mark the cylinder hole center. If an alarm lock pull no. 707 is furnished, mark the four holes on the template for the pull in addition to the cylinder hole.

3. Drill the holes as indicated for the lock and keeper. Drill the holes for the outside pull, if used.

4. Loosen the hex-head bolt holding the crossbar to the lock, and pull the bar off the clapper arm.

5. Remove the lock cover screws. Depress the latch bolt. Lift up and remove the lock cover.

6. (Disregard this step if the cylinder or thumbturn is already installed.) Remove the screws holding the cylinder housing (see Figure 13.4) to the bolt cover. Install the rim cylinder with the keyway horizontal, facing the front of the lock, and opposite the slot in the rear of the cylinder support.

7. Reinstall the support while guiding the tailpiece into the slot cam. Remove the screws holding the cylinder.

8. Use the key to try proper operation. The key should withdraw in either the fully locked or fully unlocked position of the deadbolt. If not, the cylinder and cam are mistimed. The cylinder housing must be removed, the cam turned a quarter turn to the right, and the cylinder housing reinstalled.

9. Attach the lock to the door. For single doors, remove the keeper cover and roller, install the keeper, and replace the roller and cover. Double-door keeper no. 732 is installed on the surface of the inactive leaf, as furnished.

10. Remove the hinge pivot cover. Reinstall the push-bar section on the lock clapper as far forward as it will go. Tighten the hex-head bolt under the lock clapper. Attach the hinge side pivot assembly using a level or tape to assure that the crossbar is level. If the crossbar is too long, loosen the hex-head bolt on the underside of the clapper and remove the pivot assembly. Cut the bar to the proper length, deburr the edges, and reinstall the pivot assembly on the crossbar. Only after the pivot base has been installed should the dogging screw 4 be loosened and the pivot block removed and discarded.

11. The dogging screw should face the floor. If not, remove it and reinstall it from below. Replace the pivot cover with the hole for the dogging screw facing the floor.

12. Test lock operation by projecting the deadbolt by key into the keeper. Depress the crossbar fully to retract the deadbolt, and then release the latchbolt and open the door.

13. On single doors, close the door and adjust the keeper so that the door is tightly latched. After final adjustments, install the holding screw in the keeper to maintain the position permanently. On pairs of doors, adjust the plastic slide on the keeper so that the door is tightly latched.

14. (If installing Model 35, disregard this step.) Connect the power plug. Repeat step 12. Horns should sound until the deadbolt is projected by key. Attach the self-adhesive sign to the bar and door.

15. Replace the lock cover.

Alarm Lock Models 250, 250L, 260, and 260L

Models 250, 250L, 260, and 260L by Alarm Lock Systems, Inc., are used to provide maximum security on emergency exit doors. Each of the models features sleek architectural design and finishes, a dual-piezo sounder, a low-battery alert, a simple modular construction, selectable 2-minute alarm or constant alarm, and a hardened insert in the deadbolt.

Installation

Install these models in the following way:

1. With the door closed, select the proper template. Tape it to the inside face of the door with the centerline about 38 inches above the floor, according to template directions.

2. Mark and drill the holes.

3. Remove the lock cover and four screws holding the cylinder housing to the bolt cover. Install the rim cylinder (CER) with the keyway horizontal, facing the front of the lock in the nine o'clock position. Cut the cylinder tailpiece $^3/_8$ inch beyond the base of the cylinder. Reinstall the cylinder housing, guiding the tailpiece into the cross-hole of the cam with the four screws.

4. Use the key to test for proper operation of the deadbolt. You should be able to withdraw the key from the lock in either the fully locked or fully unlocked position of the deadbolt. If not, the cylinder and the cam are misaligned, and the cylinder housing must be removed. Turn the cam one-quarter turn to the right, and reinstall the cylinder housing. *Note:* The deadbolt can be projected into the keeper by turning the key counterclockwise. Likewise, it can be withdrawn from the keeper by turning the key clockwise one full turn.

5. For outside cylinder only (CER-OKC): Install the rim cylinder with the keyway horizontal, facing the front of the door in the three o'clock position. Use the screws supplied. Cut the tailpiece $^3/_8$ inch beyond the inside face of the door.

6. For outside cylinder only: Guide the tailpiece of the outside cylinder into the cross-hole of the cam.

7. Install the lock to the door with the four screws supplied.

8. For single doors only: Remove the keeper cover, roller, and pin. Install the keeper base on the door with the two screws supplied. Reinstall the pin, roller, and cover with two screws.

9. For double doors only: Install the rub plate for a $1^3/_4$-inch-wide door from inside the door.

10. Install the keeper with two screws. Do not tighten the screws fully because the keeper will require adjustment.

11. For single doors only: Close the door, project the deadbolt, and adjust the keeper so that the door latches tightly. Retract the deadbolt, hold the keeper, release the latch, and open the door. Open the keeper cover and tighten the screws. Drill a 0.157-inch-diameter hole, as shown on the template, for the holding screw. Fasten the keeper with a screw. Reinstall the pin, roller, and cover with two screws.

12. For double doors only: Close the door, project the bolt, and adjust the plastic slide on the keeper so that the door is tightly latched. Tighten the screws.

Electronic Lock Model 265

Alarm Lock Systems Model 265 emergency exit door device has the following features: a nonhanded unit, a deadbolt with a hardened-steel insert that can be operated with an outside key, 15-second delay before the door can be opened after the clapper arm plate has been pushed, a lock that only requires 5 to 10 pounds to operate, a dual-piezo horn, and a disarming beep when the bolt is retracted with a key.

Installation

To install the Model 265 lock and keeper, proceed as follows:

1. With the door closed, select the proper template. Tape it to the inside face of the door with the centerline approximately 38 inches above the floor, according to template directions.

2. Mark and drill the following holes (see template for details):
   a. For single and double doors, mark six 0.157-inch-diameter holes, four for the lock mounting plate and two for the keeper.
   b. For double doors that are $1^3/_4$ inches thick, mark a $^1/_4$-inch-diameter hole for the rub plate on.
   c. If an outside cylinder (CER-OKC) is used, mark the center of the $1^1/_2$ inch—diameter hole.

3. If mounting the lock on a hollow metal door through which wires are run, align the lock with the holes drilled in step 2a. Mark and drill a $^3/_8$-inch hole in the door to align with the hole in the base plate near the terminal strip.

4. Remove the lock cover and four screws holding the cylinder housing to the bolt cover. Install the rim cylinder (CER) with the keyway horizontal, facing the front of the lock in the nine o'clock position. Cut the cylinder tailpiece $^3/_8$ inch beyond the base of the cylinder. Reinstall the cylinder housing, guiding the tailpiece into the cross-hole of the cam with four screws.

5. Use the key to test for proper operation of the deadbolt. You should be able to withdraw the key from the lock in either the fully locked or the fully unlocked position of the deadbolt. If not, the cylinder and cam are misaligned, and the cylinder housing must be removed. Turn the cam one-quarter turn to the right, and reinstall the cylinder housing. *Note*: The

deadbolt can be projected into the keeper by turning the key counterclock-wise one full turn.

6. For an outside cylinder only (CER-OKC): Install the rim cylinder with the keyway horizontal, facing the front of the door in the three o'clock position. Use the screws supplied. Cut the tailpiece $^3/_8$ inch beyond the inside face of the door.

7. For an outside cylinder only: Guide the tailpiece of the outside cylinder into the cross-hole of the outside cam.

8. Install the lock to the door with the four no. 10 screws that are supplied.

9. For single doors only: Remove the keeper cover, roller, and pin. Install the keeper base on the door with the two screws supplied. Reinstall the pin, roller, and cover with two screws.

10. Do not tighten the screws fully because the keeper will require adjustment.

11. For single doors only: Close the door, project the deadbolt, and adjust the keeper so that the door is tightly latched. Retract the deadbolt, hold the keeper, release the latch, and open the door.

12. Open the keeper cover and tighten the screws. Drill a 0.157-inch-diameter hole, as shown on the template, for the holding screw, and fasten the keeper with a no. 10 screw. Reinstall the pin, roller, and cover with two screws.

13. For double doors only: Close the door, project the bolt, and adjust the plastic slide on the 732 keeper so that the door is latched tightly. Tighten the screws.

14. A fine adjustment in the latch and electromagnet mechanism might be nec-essary. With the door pulled fully closed, check to see that the backstop is in complete contact with the electromagnet.

15. There will be a small gap, approximately $^1/_{32}$ inch or less, between the rod and latch. If not, loosen the Allen-head screw and slide the electromagnet to the right or left until adjusted.

16. Retighten the Allen-head screw.

Installing the control box

Install the control box for Model 265 as follows:

1. Remove the control box cover.

2. Select a location for the control box on the hinge side of the door, and mount it to the wall using the three no. $10 \times 3^1/_4$ inch self-tapping screws.

Wiring

Model 265 is wired in the following way:

1. A four-conductor no. 22 AWG cable is needed to connect the control box to the lock. There are two ways of bringing electric current from the hinge side

of the door frame to the door: Use the Armored Door Loop Model 271 and disconnect one of the 271 end boxes. Insert the loose end of the armored cable into the $\frac{1}{2}$-inch hole on the control box, and secure it with the retaining clip. Or use a continuous-conductor hinge with flying leads.

2. Connect one end of the four-conductor cable to the control box terminal strip P2 and the other end of the cable to the lock terminal strip P3. The terminal strips are marked as follows: (1) SEN—sense; (2) EM—electromagnet; (3) +9-V dc; (4) ground. *Note*: Do not cross wires.

3. Connect one end of an approved twin-lead cable to the terminal strip at Pl-1 and Pl-2, and connect the other end of the cable to the transformer provided. Do not plug in the transformer at this time.

4. Connect one end of another approved twin-lead cable to the terminal strip at Pl-3 and Pl-4. Connect the other end to the normally closed alarm relay contacts of either an approved supervised automatic fire detection system or an approved supervised automatic sprinkler system.

5. Install the control box cover.

Operating

Before installing the lock cover, do the following:

1. For a continuous alarm, leave the black jumper plug on the terminal strip installed. For 2-minute alarm shutdown, remove the black jumper plug from the terminal strip.

2. Connect the 9-V battery to its connector. A short beep will sound, ensuring that the Model 265 is powered and ready.

3. Plug the transformer into a continuous 115-V ac source. Note that the red pilot lamp on the control box is lit.

4. Open the door and install the lock cover with the four screws supplied.

Testing

To test the Model 265, lock and unlock the door using the cylinder key. Note that a small beep sounds when the bolt is retracted, indicating a disarmed condition. The door can be opened by pushing the clapper plate after a 15-second wait without causing an alarm.

Lock the door again and push the clapper plate. Immediately, the alarm will pulse loudly, and the Model 265's latch will impede opening of the door for approximately 15 seconds; then opening remains unimpeded until the lock is reset manually with the key. If continuous alarm was selected earlier, the piezo sounder will remain on for 2 minutes and then reset.

In the event of a power failure, the impeding latch will be disabled, but the 9-V battery will provide standby power for the alarm circuit. This can be tested by disconnecting the transformer from the 115-V source and pushing the clapper plate. The door should open immediately, and the alarm should sound.

In the event of a fire panel alarm, the impeding latch also will be disabled. This can be tested by disconnecting the wire going to the control box at Pl-3 and pushing the clapper plate. The door should open immediately, and the alarm should sound.

Test the unit's battery by pushing and holding the test button on the control box and pushing the clapper plate to alarm the unit. If the piezo sounder is weak or doesn't operate at all, replace the 9-V battery.

Alarm Lock Models 700, 700L, 710, and 710L

Standard features for each emergency exit door devices in Alarm Lock Systems Models 700 and 710 include a nonhanded unit, a deadbolt with a hardened-steel insert that can be operated by outside key and can be used for single or double doors, a deadlatch for easy access from inside without alarm, a loud dual-piezo horn, a selectable continuous or 2-minute alarm shutdown, a disarming beep when the bolt is retracted with a key, a low-battery beep when the battery needs to be replaced, and a retriggerable alarm after 2-minute shutdown (for Model 710 only).

Installation

Models 700 and 710 are installed as follows:

1. With the door closed, select the proper template. Tape it to the inside face of the door with the centerline approximately 38 inches above the floor, according to template directions.
2. Mark and drill the following holes (see template for details):
    a. For single and double doors, mark six 0.157-inch-diameter holes, four for the lock mounting plate and two for the keeper.
    b. Mark a $1/2$-inch-diameter hole for the rub plate on double doors $1^3/4$ inches thick.
    c. If an outside cylinder (CER-OKC) is used, mark the center of the $1^1/4$-inch-diameter holes.
    d. If an outside pull Model 707 is used, mark the center of the four 1-inch-diameter holes.
    e. If mounting the lock on a hollow metal door through which wires are run, drill hole X (see template). *Note:* If an outside pull is used, drill four $1/4$-inch-diameter holes through the door from the inside. Then drill $3/4$-inch-diameter holes $1^1/4$ inches deep from the outside of the door.
3. Remove the lock cover and four screws holding the cylinder housing to the bolt cover.
4. Install rim cylinder (CER) with the keyway horizontal, facing the front of the lock in the nine o'clock position. Cut the cylinder tailpiece $3/8$ inch beyond

the base of the cylinder. Reinstall the cylinder housing, guiding the tailpiece into the cross-hole of the cam with four screws.

5. Use the key to test for proper operation of the deadbolt. You should be able to withdraw the key from the lock in either the fully locked or the fully unlocked position of the deadbolt. If not, the cylinder and cam are misaligned, and the cylinder housing must be removed. Give the cam one-quarter turn to the right, and reinstall the cylinder housing. *Note*: The deadbolt can be projected into the keeper by turning the key counterclockwise and can be withdrawn from the keeper by turning the key clockwise one full turn.

6. For an outside cylinder only (CER-OKC): Install the rim cylinder with the keyway horizontal, facing the front of the door in the three o'clock position, using the screws supplied. Cut the tailpiece $3/8$ inch beyond the inside face of the door.

7. For an outside cylinder only: Guide the tailpiece of the outside cylinder into the cross-hole of the outside cam.

8. Install the lock loosely to the door with four no. 10 screws (supplied). Do not tighten them at this time.

9. Insert the bar and channel assembly under the channel retainer bracket, which is mounted to the lock baseplate. Hold the bar and channel assembly horizontally against the door using a level.

10. Slide the end-cap bracket into the end of the channel, and using the bracket as a template, mark and drill the two 0.157-inch-diameter mounting holes on the door. If the channel is too long, cut the channel and channel insert to the proper length and deburr the edges.

11. Attach the push bar to the lock at the clapper-arm hinge bracket using the $1/2$-inch screw and no. 10 internal tooth lockwasher provided.

12. Mount the end-cap bracket to the door with the two no. 10 screws provided, and tighten the lock securely to the door.

13. Attach the end cap to the end-cap bracket using the $1/2$-inch oval head screw provided.

14. For single doors only: Remove the keeper cover, roller, and pin. Install the keeper base on the door with the two screws supplied. Reinstall the pin, roller, and cover with two screws.

15. For double doors only: Install the rub plate for a $13/4$-inch-wide door from inside the door. Also install the Model 732 keeper with the two no. 10 screws supplied. Do not tighten the screws fully because the keeper will require adjustment.

16. For single doors only: Close the door, project the deadbolt, and adjust the keeper so that the door is latched tightly. Retract the deadbolt, hold the keeper, release the latch, and open the door.

17. Open the keeper cover and tighten the screws. Drill a 0.157-inch-diameter hole, as shown on template, for holding screw. Fasten the keeper with a no. 10 screw. Reinstall the pin, roller, and cover with two screws.

18. For double doors only: Close the door, project the bolt, and adjust the plastic slide on the Model 732 keeper so that the door is latched tightly, and tighten the screws.

Operation

Before installing the lock cover, do the following:

1. For a continuous alarm, leave the black jumper plug on the terminal strip as installed.

2. For a 2-minute automatic alarm shutdown, remove the black jumper plug from the terminal strip.

3. Connect the battery connector to the 9-V battery, observing the proper polarity. A short beep will sound, ensuring that the lock is powered and ready.

4. Install the lock cover with the four screws supplied.

5. Close the door.

Testing

To test a Model 700 or 710, do the following:

1. Lock and unlock the door using the cylinder key. Note that a small beep sounds when the deadbolt is retracted, indicating a disarmed condition. The door now can be opened without an alarm by pushing the push bar.

2. Lock the door again, and push the push bar to open it. Immediately the alarm will pulse loudly, and if continuous alarm was selected previously, the alarm will sound until the lock is reset manually by locking the deadbolt with the key. If automatic alarm shutdown was selected, the alarm will sound for 2 minutes and then reset.

When the battery becomes weak, the sounder will emit a short beep approximately once a minute, indicating that the battery needs replacing.

Special operations

If the unit has a retriggerable alarm (Model 260), after the initial 2-minute alarm and automatic shutdown, the alarm will retrigger if the door is opened again. This function will remain retriggerable until the door is relocked with the key. Whenever an alarm is caused by opening the door and the door is left open, the 2-minute alarm shutdown will be inhibited.

To use the unit's dogging operation, insert the $^3/_{16}$-inch Allen wrench into the dogging latch through the hole in the channel insert. Turn the dogging latch counterclockwise a half turn, push in the push bar, and turn the dogging latch clockwise a quarter turn until it stops. Release the push bar, and notice that it stays depressed and the door is unlatched.

Alarm Lock Model 715

Standard features of the Model 715 include a nonhanded unit, a deadbolt with a hardened-steel insert that can be operated by an outside key, a 15-second delay before the door can be opened after the push bar has been pushed, a lock that only requires 5 to 10 pounds of force to operate and can be used for single or double doors, a loud dual-piezo horn, selectable, a disarming beep when bolt is retracted with a key, and continuous or 2-minute alarm shutdown.

Installation

The Model 715 electronic exit lock is installed in the following way:

1. With the door closed, select the proper template and tape it to the inside face of the door with the centerline approximately 38 inches above the floor.
2. Mark and drill the following holes (see template for details):
   a. For single and double doors, mark six 0.157-inch-diameter holes, four for the lock mounting plate and two for the keeper.
   b. For double doors $1^3/_4$ inches thick, mark a $^1/_4$-inch-diameter hole for the rub plate.
   c. If an outside cylinder (CER-OKC) is used, mark the center of the $1^1/_4$-inch-diameter hole.
   d. If mounting the lock on a hollow metal door through which wires will be run, align the lock with the holes drilled in step.
   e. Mark and drill a $^3/_8$-inch hole in the door to align with the hole in the base plate near terminal strip P3.
3. Remove the lock cover and four screws holding the cylinder housing to the bolt cover.
4. Install the rim cylinder (CER) with the keyway horizontal, facing the front of the lock in a nine o'clock position. Cut the cylinder tailpiece $^1/_2$ inch beyond the base of the cylinder. Reinstall the cylinder housing, guiding the tailpiece into the cross-hole of the cam with four screws.
5. Use the key to test for proper operation of the deadbolt. You should be able to withdraw the key from the lock in either the fully locked or the fully unlocked position of the deadbolt. If not, the cylinder and cam are misaligned, and the cylinder housing must be removed. Turn the cam a quarter turn to the right, and reinstall the cylinder housing. *Note*: The deadbolt can be projected into the keeper by turning the key counterclockwise and can be withdrawn from the keeper by turning the key clockwise one full turn.

6. For an outside cylinder only (CER-OKC): Install the rim cylinder with the keyway horizontal, facing the front of the door in the three o'clock position. Use the screws supplied. Cut the tailpiece $^3/_8$ inch beyond the inside face of the door.

7. For an outside cylinder only: Guide the tailpiece of the outside cylinder into the cross-hole of the outside cam.

8. Install the lock loosely to the door with the four no. 10 screws supplied. Do not tighten them at this time.

9. Insert the bar and channel assembly under the channel retainer bracket, which is mounted to the lock base plate. Using a level, hold the bar and channel assembly horizontally against the door.

10. Slide the end-cap bracket into the end of the channel, and using the bracket as a template, mark and drill the two 0.157-inch-diameter mounting holes on the door. If the channel is too long, cut the channel and channel insert to the proper length, and deburr the edges.

11. Attach the push bar to the lock at the clapper arm hinge bracket using the $^1/_2$-inch screw and no. 10 internal tooth lockwasher provided.

12. Mount the end-cap bracket to the door with the two no. 10 screws provided, and tighten the lock securely to the door.

13. Attach the end cap to the end-cap bracket using the $^1/_2$-inch oval head screw provided.

14. For single doors only: Remove the keeper cover, roller, and pin. Install the keeper base on the door with the two screws supplied. Reinstall the pin, roller, and cover with two screws.

15. For double doors only: Install the rub plate for a $1^3/_4$-inch wide door from inside the door. Also install the Model 732 keeper with the two no. 10 screws supplied. Do not tighten the screws fully because the keeper will require adjustment, as mentioned in the next step.

16. For single doors only: Close the door, project the deadbolt, and adjust the keeper so that the door is latched tightly. Retract the deadbolt, hold the keeper, release the latch, and open the door.

17. Open the keeper cover and tighten the screws. Drill a 0.157-inch-diameter hole, as shown on the template, for the holding screw, and fasten the keeper with a no. 10 screw. Reinstall the pin, roller, and cover with two screws.

18. For double doors only: Close the door, project the bolt, and adjust the plastic slide on the Model 732 keeper so that the door is latched tightly. Tighten the screws.

Installing the control box

To install the control box for Model 715, remove the control box cover. Select a location for the control box on the hinge side of the door, and mount it to the wall using the three $^3/_4$-inch self-tapping screws.

Wiring

The Model 715 is wired in the following way:

1. A four-conductor no. 22 AWG cable is needed to connect the control box to the lock. There are two ways of bringing electric current from the hinge side of the door frame to the door. Use the Armored Door Loop Model 271 by disconnecting one of the Model 271 end boxes and inserting the loose end of the armored cable into the $1/2$-inch hole on the control box. Secure it with the retaining clip. Or use a continuous-conductor hinge with flying leads.

2. Connect one end of the four-conductor cable to the control box terminal strip P2 and the other end of the cable to the lock terminal strip P3. The terminal strips are marked as follows: (1) SEN—sense; (2) EM—electromagnet; (3) +9-V dc; and (4) ground. *Note*: Do not cross wires.

3. Connect one end of an approved twin-lead 18-2 cable to the terminal strip at Pl-1 and Pl-2, and connect the other end of the cable to the 12-V ac, 20-VA transformer provided. Do not plug in the transformer at this time.

4. Connect one end of another approved twin-lead 22-2 cable to the terminal strip at Pl-3 and Pl-4, and connect the other end to the normally closed alarm relay contacts of either an approved supervised automatic fire detection system or an approved supervised automatic sprinkler system.

5. Install the control box cover.

Operating

Before installing the lock cover, do the following:

1. For a continuous alarm, leave the black jumper plug on the terminal strip as installed.

2. For a 2-minute automatic alarm shutdown, remove the black jumper plug from the terminal strip.

3. Connect the 9-V battery to its connector. A short beep will sound, ensuring that the model 715 is powered and ready.

4. Plug the transformer into a continuous 115-V ac source. Note that the red pilot lamp on the control box is lit. Open the door and install the lock cover with the four screws supplied.

Testing

To test a Model 715, do the following:

1. Lock and unlock the door using the cylinder key. Notice the small beep when the deadbolt is retracted, indicating a disarmed condition. The door can be opened by pushing the push bar after a 15-second wait without causing an alarm.

2. Lock the door again, and push the push bar. Immediately, the alarm will pulse loudly, and the Model 715 latch will impede opening of the door for approximately 15 seconds and then remain unimpeded until it is reset manually with the key. Or, if a continuous alarm was selected above, the piezo sounder will remain on for 2 minutes and then reset.

3. In the event of a power failure, the impeding latch will be disabled, but the 9-V battery will provide standby power for the alarm circuit. Test this by disconnecting the transformer from the 115-V source and pushing the push bar. The door now should open immediately, and the alarm should sound.

4. In the event of a fire alarm, the impeding latch also will be disabled. Test this by disconnecting the wire going to the control box Pl-3 and pushing the push bar. The door should open immediately, and the alarm should sound.

5. Test the battery by pushing and holding the test button on the control box and pushing the push bar to alarm the unit. If the piezo sounder sounds weak or doesn't operate at all, replace the 9-V battery.

4

Securing Windows

People who do a lot to secure their doors may be paying little attention to their windows because they think that securing windows is time consuming, expensive, or impossible. To burglars, windows are often the most attractive entry points.

The materials used in making doors are also used for manufacturing window frames. Wood, aluminum, fiberglass, and polyvinyl chloride (PVC) plastic are most popular for windows. As long as the windows are well built and have good locking devices (keyless types are best), the frame material usually has little effect on a home's security.

Contrary to popular opinion, it usually isn't necessary to make your window frames and panes unbreakable to keep burglars out—unless your neighbors are out of earshot. Burglars know that few things attract more attention than the sound of breaking glass, and they don't like to climb through openings that have large jagged shards of glass pointing at them. When they can't get into a house without breaking a window, most burglars will move on to another house.

You can make your windows more secure just by making them hard to open quietly from outside. Don't install a lock or any other device that might delay a quick exit in case of a fire. Balancing the safety and security elements depends on what type of windows you have. The four basic types are sliding, casement, louvered, and double hung.

A sliding window works much like a sliding-glass door, and like a sliding-glass door, it usually comes with a weak lock that's easy to defeat. Most of the supplemental locking devices available for sliding windows fit along the track rail and are secured with a thumbscrew. You then can keep the window in a closed or a ventilating position depending on where you place the thumbscrew. The need to twist a thumbscrew can be inconvenient if you must lock and unlock a window frequently.

Another keyless device for securing sliding windows is offered by M.A.G. Manufacturing, Inc. The company's Quick-Vent Model 8830 is a heavy-gauge-metal locking device that allows you to secure a window in three different positions. The device is held in place by security set screws instead of thumb-screws. Quick-Vent can be purchased for less than $20.

A casement window is hinged on one side and swings outward (much like your doors do). It uses a crank or handle for opening and closing. To prevent some-one from breaking the glass and turning the crank, the handle should be removed when it isn't being used.

Louvered (or *jalousie*) windows are the most vulnerable type. They are made of a ladder-like configuration of narrow, overlapping slats of glass that can be pulled out of the thin metal channels easily. Jalousies attract the attention of burglars and should be replaced with another type of window.

The type of window used in most homes is the double-hung window. It con-sists of two square or rectangular sashes that slide up and down and are secured with a metal thumb-turn butterfly sash "lock." (Although most manufacturers call it a lock, the device is really just a clamp.) The device holds the sashes together in the closed position, but a burglar can work it open by shoving a knife in the crack between the frames.

Watchguard, Inc., makes a useful replacement for conventional sash locks. The company's Safety Sash Lock can't be opened from the outside of a home. It looks like a standard sash lock but incorporates a spring-loaded lever that prevents it from being manipulated out of the locked position. The Safety Sash Lock sells for about $10.

As an alternative to replacing sash locks, a ventilating wood window lock can be installed. This device allows someone inside to raise the window a few inches and then set the bolt, which prevents anyone outside from raising the window higher. The device consists of an L-shaped metal bolt assembly and a small metal base. The bolt assembly fits along the inner edge of either of the two stiles (vertical members) of the top sash and is held in place with two small screws. The base is attached, in alignment, on the top of the bottom sash and is held in place with one small screw.

The bolt assembly has a horizontal channel that allows you to slide the bolt into the locked and unlocked positions. When in the locked position, the bolt extends over the base and obstructs the bottom sash from being raised past the bolt. When in the unlocked position, the bolt is parallel to the window and out of the way of the bottom sash. The higher you place the bolt mechanism above the bottom sash, the higher you'll be able to raise the window with the bolt in the locked position. The base isn't really needed, but it helps prevent the bottom sash from getting marred.

Many companies make ventilating wood window locks, and there aren't any important differences between brands. Most models are sold at home-improvement centers and hardware stores for less than $5 each.

Glazing

Glazing is a term that refers to any transparent or translucent material—usually some kind of glass or plastic—used on windows or doors to let in light.

Most types of windows can be made more secure by replacing the glazing with more break-resistant material.

The most common glazing for small windows is standard sheet glass. It's inexpensive, but it can easily splinter into small, sharp pieces. Plate glass, which is a little stronger than sheet glass, generally is used in large picture windows. Because plate glass also has the problem of breaking into many dangerous pieces, it shouldn't be used in exterior doors.

Tempered glass is several times stronger than plate glass and costs about twice as much. Rather than shattering into many sharp pieces, tempered glass breaks into small, harmless pieces—the reason for using it in patio doors. When a large piece of tempered glass breaks, it makes a lot of noise, which may attract the attention of neighbors.

The strongest type of glass a homeowner might use is laminated glass: It's made of two or more sheets of glass with a plastic inner layer sandwiched between them. The more layers of glass and plastic, the stronger (and more costly) the laminated glass will be. Laminated glass 4 inches thick can stop bullets and is often used for commercial applications.

Plastics are used commonly as glazing materials. Acrylics such as Plexiglas and Lucite are very popular because they are clearer and stronger than sheet glass. However, they scratch easily and can be sawed through. The strongest types of plastic that a homeowner might use are polycarbonates such as Lexigard and Lexan. Although they're not as clear as acrylics, polycarbonates are up to 30 times stronger. Untreated polycarbonates scratch easily, but you can buy sheets with scratch-resistant coatings.

For increased strength and energy efficiency, many modern windows come in parallel double- or triple-pane configurations. Three parallel panes can provide good security.

Whether you want to replace your glass panes with stronger glass or replace broken panes, you can easily do it yourself. In addition to being unsightly, a broken window can attract burglars because they can quietly remove the glass to gain entry. A broken pane should be replaced immediately.

Glass Blocks

Glass blocks come in a wide variety of patterns and sizes and are strong enough to be used in place of plate glass. They are especially useful for securing basement windows. Most patterns create a distorted image for anyone trying to see through them, but clear glass blocks are also available.

For areas that require ventilation, you can buy preassembled panels of glass block with built-in openings. Preassembled panels are easy to install if you get the right sizes. To order the right size panel, you need to know the size of your window's rough opening. If you have a wood-frame wall, you can determine the rough opening by measuring the width of the opening between the frame's sides and the height between the sill and the header. You'll need a panel about $1/2$ inch smaller than that measurement to make sure that it will fit in easily.

If you have a masonry wall, you can determine the rough opening by measuring the width between the brick or block sides and the height between the header and the sill. Be sure that the panel you order is about $1/2$ inch smaller than the opening.

There are two basic ways to install glass block panels. The older way involves using masonry cement or mortar—in much the same way as when installing bricks. The newer way involves using plastic strips.

A newer way to install glass blocks

A cleaner and simpler way to install glass blocks was developed recently by Pittsburgh Corning. However, the company's glass-block panel kits aren't designed to offer strong resistance to break-in attempts. The kit includes: U-shaped strips of plastic channel for the perimeter of the panel, a roll of clear plastic spacer for holding the blocks in place, clear silicon caulk, and a joint-cleaning tool. The kit can be used in the following way:

1. Install the U-shaped channels, using shims if necessary. Attach the channels with 1-inch flathead wood screws, and conceal the screw heads with white paint.

2. Using short lengths of spacer, align the blocks vertically. (Use a utility knife to cut spacers to size.)

3. Align the blocks horizontally with long strips of spacer. (The spacer's contour matches the block edge surface.)

4. Fit the last block into the panel through a section removed temporarily from the top channel.

5. Slip the leftover section of the top channel over the last block and apply caulk to hold the section in place.

6. Wipe the joints clean with a cloth dampened with isopropyl alcohol; fill the joints with beads of silicon caulk.

Protecting Glass

If you're concerned that someone will gain entry by breaking a window, you might want to coat the glass with security film, a transparent laminated coating that resists penetration and holds broken glass firmly in place. Even after breaking the glass, burglars would have a hard time getting through. Some security film can hold glass in place against sledgehammer blows, high winds, and explosions.

A less aesthetic but equally effective way to protect windows is to install iron security bars. The mounting bolts should be reachable only from inside your home. Be sure the bars don't make it hard for you to escape quickly if necessary. Hinge kits are available for many window bars.

Before buying window bars, you'll need to measure the width and height of the area to be covered. In general, the larger the area to be protected, the more the bars will cost. Window bars range in price from about $10 to more than $50 per window.

How to Secure a Double hung Window at No Cost

1. From inside the home, close the window, and clamp the butterfly twist-turn sash lock into the closed position.

2. Use a pencil to mark two spots below the twist-turn lock on the top rail (horizontal member) of the bottom sash. One mark should be about 1 inch inside the left stile; the other should be about 1 inch inside the right stile.

3. Position your drill at the first mark, and drill at a slightly downward angle until your drill bit goes completely through the top rail of the bottom sash and about halfway through the bottom rail of the top sash. Then do the same thing at the other mark you made.

4. Raise the bottom sash about 5 inches, and hold it steady. Insert the drill bit back into one of the bottom sash holes, and drill another hole about halfway through the top sash. (The hole should be about 5 inches above another hole you drilled on the stile.) Without moving the sash, do the same thing at the other side of the window.

5. Close the window, and insert two small nails or eye bolts into the lower sets of holes to hold the sashes together so that the window can't be lifted open from outside. When you want ventilation, you can remove the nails or bolts, raise the window, and insert them in the top set of holes to secure the window in the open position.

5

Lock Basics

Laypersons frequently use a generic name such as *padlock, automobile lock*, or *cabinet lock* when referring to a lock. Such a name has limited value to locksmiths and other professionals who install and recommend locks because it is too general. It simply refers to a broad category of locks that are used for a similar purpose, share a similar feature, or look similar to one another.

Locksmiths identify a lock in ways that convey information needed to purchase, install, and service it. The name they use is based not only on the purpose and appearance of the lock but also on the lock's manufacturer, key type, method of installation, type of internal construction, and function.

The names used by a locksmith typically are formed by combining several words. Each word in the name provides important information about the lock. The number of words a locksmith uses for a name depends on how much information he or she needs to convey.

When ordering a lock, for instance, the locksmith needs to use a name that identifies the lock's purpose, manufacturer, key type, appearance, etc. However, a name that simply identifies the lock's internal construction may be adequate for describing a servicing technique to another locksmith.

Grades of Door Locks

Most lock manufacturers offer locks in several grades, such as *light duty, residential, heavy duty*, and *commercial*. Such descriptions are useful guidelines, but there are no industry standards for manufacturing each grade. One manufacturer's *heavy duty*, for example, may be of lower quality and less resistant to break-ins than another manufacturer's *light duty*. The grade names are meaningful only for comparing locks made by the same manufacturer.

A better measure of a lock's quality is the rating given to it by the American National Standards Institute (ANSI). The three most common ANSI standards for locks—grade 1, grade 2, and grade 3—are based on a range of performance features, especially during impact tests. The ANSI tests gauge how well the locks resist forceful attacks and whether their finishes hold up well over extended periods.

When you see one of these ANSI ratings, this is what you should know:

Grade 1 locks are for heavy-duty commercial uses and would be overkill for most homes.

Grade 2 locks—although designed for light commercial uses—can provide good protection for most homes.

Grade 3 locks are good for light residential applications.

Most door locks sold to homeowners are either grade 3 or have no ANSI classification.

Only a few lock manufacturers—Kwikset Corporation and Master Lock Company, for example, offer lines of grade 2 locks through locksmith shops, department stores, and home-improvement centers. These ANSI-graded locks often cost little more than locks that have no ANSI classification.

Should you purchase a lock that does not have an ANSI classification? Aside from the directions on how to install it, you can't rely on any information the manufacturer has printed on the packaging. Much of the description of what the lock can do for you is just advertising hype. To be able to separate the hype from meaningful information, you'll need to know the strengths and weaknesses of the generic types of door locks.

Types of Door Locks

Six types of locks are used commonly to secure doors in homes:

1. Bit-key
2. Key-in-knob
3. Key-in-lever
4. Deadbolt
5. Jimmy-proof deadlock
6. Multiple-bolt

You may not be familiar with their names, but you've probably seen all six types of locks. The first three are not recommended as sole exterior door locks.

The bit-key lock, the oldest type, has a large keyhole and works with a "skeleton key." A bit-key lock (Figure 5.1) is all right for closet or bathroom doors but shouldn't be used on exterior doors because it's very easy to defeat. Many people know how to unlock a bit-key lock with a metal coat hanger, and anyone who buys the stock bit-key locks sold at many hardware stores can unlock most any bit-key lock.

Figure 5.1 Bit-key locks often are used on bathroom and closet doors. (*Courtesy of M.A.G. Manufacturing.*)

The most popular exterior door lock installed in homes built since the late 1950s is the key-in-knob lock (Figure 5.2). Basically, the key-in-knob lock is two connecting doorknobs with a keyway in one doorknob or in both. The key-in-knob lock is inexpensive and easy to install. It allows the door to be locked or latched—without a key—just by closing the door.

Like a bit-key lock, a key-in-knob lock is easy to defeat. It can't be opened with over-the-counter keys or coat hangers, but the key-in-knob lock is vulnerable in

Figure 5.2   A key-in-knob lock.

Figure 5.3   A lever-handle lock.

other ways. After hammering one of the knobs off, for instance, a burglar can use a screwdriver to retract the lock's bolt from the door frame. There are also quieter ways to defeat the key-in-knob lock. Like the bit-key lock, the key-in-knob is fine for a bathroom or closet door but shouldn't be used as the sole lock for an exterior door.

Internally, the key-in-lever or lever-handle lock (Figure 5.3) is designed a lot like the key-in-knob lock. Both have the same vulnerabilities. The key-in-lever lock consists of one or two levers instead of knobs. The levers make it easier for young children and people with physical handicaps to use these locks. A knob-and-lever-handle lock has both a knob and a lever handle (Figure 5.4).

If your door has one of the three types of locks just described, it may be a good idea for you to gain extra security by installing a stronger lock above your present key-in-lever, key-in-knob, or bit-key lock. You might want to use a deadbolt lock (Figures 5.5 and 5.6), a jimmy-proof deadlock (Figure 5.7), or a multiple-bolt lock. Each is described and illustrated in the sections that follow.

Choosing a Deadbolt

Many police departments recommend using a tubular deadbolt (or deadbolt, for short) on exterior doors. However, some deadbolts aren't much better than a key-in-knob lock. To understand why, you have to know how deadbolts work and how burglars try to defeat them.

The main parts of a typical deadbolt include a cylinder (Figure 5.8), cylinder guard, bolt assembly, tailpiece, thumbturn, and mounting screws. The cylinder is the cylindrical metal part with a keyway on its face, and the cylinder guard (or collar) fits around the cylinder. Both fit together and are mounted on the exterior side of the door. Using the mounting screws, the thumbturn is mounted behind the cylinder on the interior side of the door. The bolt assembly and tailpiece fit within a cavity in the door, between the thumbturn and the cylinder. It is the tailpiece that connects the thumbturn and cylinder to the bolt assembly.

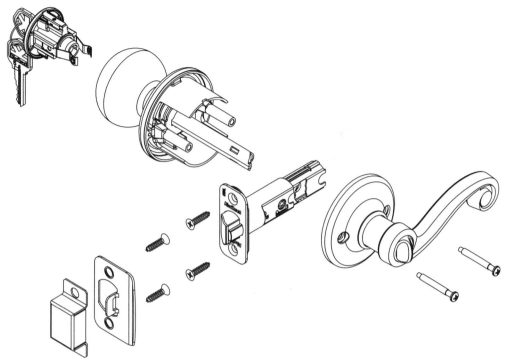

Figure 5.4 A knob-and-lever-handle lock.

Here's how a deadbolt works. On the exterior side of the door, when you turn a key in the cylinder, the connecting tailpiece engages the bolt mechanism, which causes the bolt to extend or retract. On the interior side, when you twist the thumbturn, the connecting tailpiece engages the bolt mechanism and causes the bolt to extend or retract.

Some high-security deadbolts have special parts. The Abloy Disklock, for example, includes adapter rings and a bolt assembly protector. The bolt assembly protector fits between the cylinder and the bolt assembly and has a circular

Figure 5.5 A deadbolt lock. (*Courtesy of Medeco High Security Locks.*)

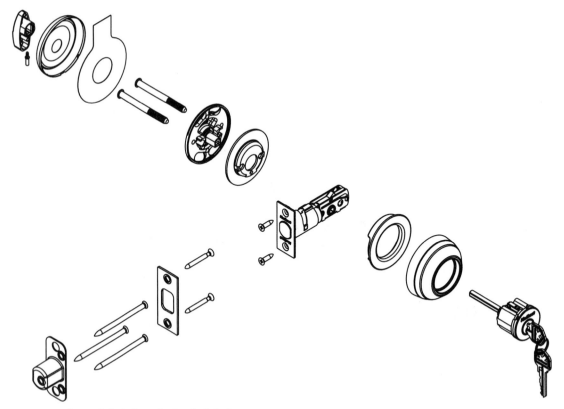

Figure 5.6   An exploded view of a deadbolt lock.

opening to allow the tailpiece to be connected to the cylinder. Because it has a hood that covers the internal parts of the bolt assembly, the bolt assembly protector prevents a burglar from manipulating the bolt assembly with an ice pick or other instrument. Adapter rings are used to make a lock fit better on a thin door. They are installed between the door and the cylinder guards.

Figure 5.7   A jimmy-proof dead-lock.   (*Courtesy   of   M.A.G. Manufacturing.*)

Figure 5.8 A lock cylinder. (*Courtesy of Medeco High Security Locks.*)

How deadbolts are defeated and what you can do about it

Common methods of attacking deadbolts include jimmying, sawing, wrenching, and lock picking.

Jimmying. Jimmying is done by inserting a pry bar between a door and its frame, near the extended bolt, and prying until the bolt is freed from its strike place. The longer the bolt, the harder the lock will be to jimmy. Make sure that your deadbolt has a 1-inch throw: 1 inch of the bolt should extend past the edge of the door.

Sawing. If the bolt is made of a soft metal, such as brass, it can be sawed off with a hacksaw blade. Make sure that your lock's bolt is made of steel or has a hardened-steel insert.

Wrenching. When a wrench is clamped onto a cylinder guard, the cylinder can be twisted off the door. Only use deadbolts that come with tapered free-spinning cylinder guards. They're hard to wrench because they just spin around the cylinder rather than turning it.

Lock picking. In theory, any lock that uses a key can be picked. It's true that it's always *possible* to create an instrument that can be used to simulate the action of a key, but most people don't need to worry that a burglar will pick open their locks. Contrary to what's portrayed in movies, few home burglars pick open door locks. Lock picking is a sophisticated skill that takes a long time to learn, and there are usually faster and easier ways to break into a home.

Suppose that you lost your last set of house keys and chose not to break a window or glass panel to gain entry into your home. An experienced locksmith may need 10 or more minutes to pick open your exterior door lock. This professional would be intent on opening the door without damaging the lock or the door. Burglars are concerned about gaining entry quickly, and they care little about the damage they might cause. Why would burglars spend 5 to 10 frustrating minutes fiddling with lock picking when they can kick the door in or jimmy the lock instead?

If you live in a large city or near a high-crime area, however, you may need to be concerned about lock picking. Such places have more than their share of sophisticated burglars who can pick open locks faster than many locksmiths can. The best way to thwart lock-picking attempts is to use high-security cylinders in the locks on all your exterior doors. (Information about high-security cylinders is given later in this chapter.)

Installing a deadbolt

A deadbolt lock is installed by drilling two holes: a larger hole through the face and back of the door and a smaller hole through the edge of the door. The larger hole is usually either $2^3/_4$ or $2^3/_8$ inches in diameter. The smaller hole is usually 1 inch in diameter or smaller. It's needed for the bolt to go through. The distance from the door edge to the center of the large hole is called the *backset*. Most backsets are either $2^3/_8$ or $2^3/_4$ inches.

You need to know your door's exact measured distances before you buy or install any door lock. For a new installation, you need the measurements to choose the right size drill bits. For a retrofit installation, you need the measurements to choose a lock that lets you use the preexisting installation holes.

Most deadbolts come with a template. Wrap the template around the edge of the door, and mark on the door the spots where you are to drill the holes. Be sure the template is straight, and place your marks on the high side of the door bevel. (The lock edge of your door is slightly beveled so that the door can open and close easily.)

When drilling a lock hole through a door, always use a pilot drill bit that extends at least 1 inch past your hole saw. Begin drilling on the high side of the door. When your pilot bit penetrates through the other side of the door, stop drilling. Go to the side of the door that your pilot bit punctured, and use the small puncture hole as a guide to finish drilling the lock hole. By working on both sides of the door when you're drilling a lock hole, you'll be less likely to splinter the door. Be sure to *drill straight*. If you have a hard time drilling straight, you might want to attach a small level onto your drill or use a lock hole-boring jig.

If you're installing a deadbolt on a metal door or a hollow-core wood door, you'll need to install a lock-support insert in the door to prevent the lock from loosening. Lock-support inserts generally cost less than $10.

Choosing a Jimmy-Proof Deadlock

A jimmy-proof deadlock (or vertical deadbolt) can offer a lot of resistance to jimmying attempts. It has a rectangular body with two or three cylindrical bolts at one end. The lock is surface mounted on a door so that its bolts are aligned with the "eyeloops" of a matching strike plate. When placed in the locked position, the bolts drop vertically into the eyeloops, and the lock can't be separated from the strike plate without breaking the entire unit.

One problem with a jimmy-proof deadlock is that it's effective only when installed properly on a strong door frame and door. In many homes, the door frames aren't strong enough to support a jimmy-proof deadlock properly. Another problem is that if the mounting screws for the strike plate aren't long enough, the strike plate won't hold to the frame during a kick-in. Be sure to use screws that are at least 3 inches long when you install a strike plate.

In jimmy-proof deadlocks, unlike deadbolts, the cylinders aren't protected by cylinder guards, which makes jimmy-proof deadlocks vulnerable to drilling, pulling, and chiseling attacks. (These sophisticated types of attacks rarely occur outside large cities and high-crime areas.) The easiest way to thwart these attacks is to install a hardened-steel cylinder guard plate over the jimmy-proof deadlock. The guard plate covers virtually every vulnerable area of the cylinder, leaving only the keyway exposed.

Choosing Multiple-Bolt Locks

For maximum lock protection, you can use a multiple-bolt lock, a type of lock that has two or more long bolts that operate simultaneously. The bolts may extend vertically, horizontally, or in several directions. Usually, a key unlocks the door from the outside, and a thumbturn moves the bolt from the inside. There are two basic kinds of multiple-bolt locks: surface mount and mortise.

The surface-mount type has two or more bolts that extend and retract across a door's surface. A popular surface-mount style, often called a *police lock*, has two bolts: One extends horizontally into the hinged side of a door, and the other extends into the door's opening side.

The mortise type of multiple-bolt lock has bolts that move within a hollowed-out cavity in the door. These locks can be used only on a thick wood or metal door. They usually need to be installed by a professional who has the special tools needed to prepare the door.

Lock Cylinders

Regardless of the type of lock you choose, you'll need to decide whether you want a single- or double-cylinder model. A single-cylinder lock is one that uses a key on one side of the door and a thumbturn on the other. A double-cylinder lock requires that a key be used on both sides of the door.

Security professionals disagree about which type of lock cylinder is best. Some point out that double-cylinder locks force a burglar who climbed in through a window to leave through a window—which lessens the amount that can be carried out. For this strategy to work, however, you would need to use double-cylinder locks on every exterior door in your home, and you would have to install all of them in a way that would make removal difficult.

I don't recommend this strategy: It would present a serious problem if you needed to get out quickly—such as during a fire—and couldn't find the key. Your door locks should be for keeping unwanted people outside, not for trapping yourself inside.

Some lock makers, such as Abloy High Security Locks and Kwikset, offer double-cylinder locks with safeguards. These devices allow the locks to work like single-cylinder models while people are inside a home. Never use double-cylinder locks that lack this safety feature for all the exterior doors of your home.

High-security cylinders

Any lock can be made stronger by replacing its standard cylinder with a high-security cylinder. A high-security cylinder provides special protection against lock picking, drilling, and other entry techniques that sophisticated burglars use. Although any lock that requires a key can be picked open eventually, locks with high-security cylinders are almost never defeated by lock pickers.

Ordinary lock picks are seldom successful on a high-security cylinder. Even with custom-made tools, it could take a professional burglar (or a locksmith) several hours to pick open a typical high-security cylinder. Few home burglars are willing to spend that much time at one place.

An important advantage in using a high-security cylinder is that you'll have maximum control over who may get a copy of your key. The easier a key can be duplicated, the easier a lock can be compromised. If you're like most people, you've entrusted your house keys at one time or another to friends, neighbors, a home-repair contractor who needed access at a time when you could not be home, a parking-lot attendant (if you keep all your keys on the same key ring), and others. With standard lock cylinders, you never know when someone might make copies of your keys at the nearest hardware store.

Keys for high-security cylinders are hard for unauthorized persons to duplicate because they can be copied only on special key machines. Manufacturers of high-security cylinders often use patented keys to further ensure key control. Patented keys can be duplicated only by a select group of locksmiths who have been approved by the manufacturers and who will require proof of lock ownership from anyone wishing to duplicate the keys.

There are many brands of high-security cylinders; most are sold only through locksmiths. Because each line involves a large initial investment for merchandise, special machines, special tools, training, and the like, few locksmiths carry more than two or three high-security product lines. Don't be surprised if the locksmith you approach tries to explain in great detail why the one line he carries is unique and superior to all others.

The major high-security cylinders—those made by Abloy, Assa, Medeco, and Schlage Primus, to name a few—have unique, patented features. Your only concern should be whether the cylinder is listed by Underwriters Laboratories (UL), an independent testing agency. Any UL-listed model easily can meet or exceed the needs of most homeowners. A UL listing means that a sample model has withstood rigorous expert attack tests.

You can buy a high-security cylinder for under $100. Make sure that the one you select fits the lock it is intended for. Find out how much you'll have to pay for duplicate keys. Depending on the brand, you can expect to pay between

$3 and $10 per key. See Chapter 6 for more information about high-security locks and cylinders.

Making a standard cylinder more secure

For about $10, you can make a standard cylinder "almost high security." Just have a locksmith rekey the cylinder using two mushroom or spool pins, and the cylinder will be more pick-resistant. Ask the locksmith to make your new keys on high-security key bow (rhymes with "toe") blanks. This type of key blank doesn't show the information that most key cutters need to duplicate keys. Only highly experienced locksmiths know how to duplicate a key made on a high-security key bow blank. The fewer people who know how to duplicate your key, the harder it will be for someone to quickly have a duplicate made.

Minimizing Your Keys

Four thousand years after door keys were invented, we still rely on them for security. Unlike the large wooden keys the early Egyptians proudly carried on their shoulders, today's small metal keys are often more a burden than a source of pride. They tear holes in pockets, they're inconvenient to carry around, and they're easy for children and adults to lose. If you're like most people, you have to fumble through eight or more keys to get into your home. There are simple ways to reduce your need for keys—without reducing your security.

One way to make your key ring smaller is to have all or most of your locks work with the same key. Some people have several door locks that use different keys because they think burglars have more trouble picking open a variety of locks. Locks that are keyed alike are no easier to defeat than those that are keyed differently. They still have to be picked open one at a time.

If you want the convenience of using one key for your doors, some of your locks may need to be rekeyed (so that all of them have the same tumbler pattern). Locksmiths charge about $10 per lock for rekeying.

Sometimes it isn't practical to make two locks fit the same key. If the key to one can't slide into the keyway of another, for example, then those two locks can't share a key. Locks come in hundreds of different keyway shapes and sizes, and each can accept only certain keys.

You can avoid the hassle and extra costs of rekeying if you consider key compatibilities when you're buying new locks. Many locks sold at hardware and department stores have keying numbers (or tumbler-pattern numbers) on their packages. You may be able to find several locks with the same number, which means that they're keyed alike. (Because there are so many possible keying numbers, when you see the same numbers on more than one package, you can assume that those locks also have the same key way.)

If you use padlocks around your home, you may want to get models that will work with your door key. You can find them at locksmith shops. Bring your door key with you to make sure that the padlock has the right keyway, and have it rekeyed to match your door lock.

The exact cost will depend on the size, type, and brand of lock you choose, but expect to pay at least $20 for any padlock that uses your door key. If you don't want to spend that much, you can buy a good combination padlock for much less. (If the body and shackle of a combination padlock are strong, it can provide as much protection as a key-operated padlock.)

A combination padlock isn't the only keyless lock you can use in your home. Several companies make pushbutton locks for exterior doors. These locks are especially useful for young children or anyone who has a hard time keeping track of keys. Make sure that any keyless lock you install on an exterior door is weather resistant and has a rigid bolt that projects from the edge of the door. Stay away from exterior door locks with spring-loaded bolts only. They're easy to defeat. (If the bolt is beveled, you can assume it's spring-loaded.)

Simplex Mechanical Pushbutton Locks

Simplex mechanical pushbutton locks offer a convenient way to control access between public and private areas (Figure 5.9). There are no keys or cards to manage, no computers to program, no batteries to replace, and combinations can be changed in seconds without removing the lock from the door.

Figure 5.9 A Simplex pushbutton lock. (*Courtesy of Simplex.*)

6

High-Security Mechanical Locks

With respect to locking devices, the term *high security* has no precise meaning. Some manufacturers take advantage of this fact by arbitrarily using the term to promote their standard locks.

Most locksmiths would agree that to be considered "high security," a lock should have features that offer more than ordinary resistance against picking, impressioning, drilling, wrenching, and other common burglary techniques. The most secure locks also provide a high level of key control. The harder it is for an unauthorized person to have a duplicate made, the more security the lock provides.

Underwriters Laboratories' Listing

Founded in 1894, Underwriters Laboratories, Inc. (UL), is an independent non-profit product-testing organization. A UL listing (based on UL Standard 437) is a good indication that a door-lock cylinder offers a high degree of security. If a lock or cylinder has such a listing, you'll see the UL symbol on its packaging or on the face of the cylinder.

To earn a UL listing, a lock or cylinder must meet strict construction guidelines, and a sample model must pass rigorous performance and attack tests. Some of the requirements are as follows:

- All working parts of the mechanism must be constructed of brass, bronze, stainless steel, or equivalent corrosion-resistant materials or have a protective finish complying with UL's salt-spray corrosion test.

- All locks must have at least 1000 key changes.

- All locks must operate as intended during 10,000 complete cycles of operation at a rate not exceeding 50 cycles per minute.

- All locks must not open or be compromised as a result of attack tests using hammers, chisels, screwdrivers, jaw-gripping wrenches, pliers, hand-held electric drills, saws, puller mechanisms, key-impressioning tools, and picking tools.

The attack test includes 10 minutes of picking, 10 minutes of key impressionng, 5 minutes of drilling, 5 minutes of sawing, 5 minutes of prying, 5 minutes of pulling, and 5 minutes of driving. These are networking times, which don't include time used for inserting drill bits or otherwise preparing tools.

Key Control

Another important factor in lock security is *key control*. The most secure locks have patented key blanks that can be cut only on special key machines. This type of key control greatly reduces the number of places where an unauthorized person can have a key duplicated. The least secure locks use keys that can be copied at virtually any hardware or department store.

Lock and Key Patents

To increase key control, a lock and/or its key blanks are patented. This makes it harder to have an unauthorized key made.

Not all patents offer a lot of key control, however. There are two relevant types of patents: design and utility. A design patent only protects how a thing looks. In 1935, Walter Schlage received a design patent for a key bow. The bow's distinctive shape made it easy for locksmiths and key cutters to recognize the company's keys. But the patent didn't prevent aftermarket key-blank manufacturers from making basically the same key blanks with slightly different bow designs. Design patents are good for 14 years.

Utility patents are more popular and provide more protection against unauthorized manufacturing. All patents expire eventually, and they aren't renewable. Utility patents are granted for 20 years, beginning on the application filing date. Medeco Security Lock's original patent prevented companies from making key blanks that fit Medeco locks. The company was able to maintain maximum key control. As soon as the patent expired (in 1986), many companies began offering compatible key blanks. The company responded by introducing a new lock, the Medeco Biaxial, getting a utility patent (No. 4,393,673) that expired in 2004. The company's newest patented high-security lock is its Medeco 3, whose patent expires in 2021.

Just because a patent has expired doesn't mean that there are aftermarket blanks available for the lock. High-security lock makers plan for patent expirations. They make many different keyways, and keep that number a secret. They also make a point of evenly distributing keyways to different geographic areas. Aftermarket blanks are made only if the blank makers can sell enough blanks to make it worth their while. If you're thinking about carrying a high-security line, ask the manufacturer how many keyways it has. A good manufacturer won't tell you.

Types of High-Security Mechanical Locks

The rest of this chapter provides detailed information about some of the most popular high-security mechanical locks. Much of the information was obtained from manufacturers' technical bulletins and service manuals.

Mul-T-Lock high-security system

Mul-T-Lock high-security cylinders have a unique telescopic pin-tumbler mechanism with internal and external pins. Both the internal and the external shear lines must be aligned simultaneously in order for the plug to rotate.

The Mul-T-Lock patented plug has a unique structure. When the top and bottom pins, plug, and body meet, a three-dimensional shear line is formed to an almost perfectly spherical shape. Steel inserts ensure antidrilling resistance.

These features provide an added security dimension and make the Mul-T-Lock cylinder pick resistant and drill resistant for high-security needs. When master keyed, additional side pins or back pins can be incorporated.

The Mul-T-Lock patented Interactive system raises the level of security to new heights. The Interactive system combines the unique telescopic pin-tumbler mechanism and special features of the Classic system with a spring-loaded pin in the cylinder plug to produce a virtual combination only when the key is inserted in the lock.

The Interactive patented key and key blank provide increased control over key cutting to achieve an even higher level of key security. Additional keys should be cut only after presentation of a Mul-T-Lock key card and verification of customer identity in accordance with Mul-T-Lock key-cutting procedures, which may be obtained through Mul-T-Lock professional locksmiths. Interactive technology is retrocompatible with the Mul-T-Lock Classic system, allowing existing locks to be upgraded.

Mul-T-Lock Integrator system key security

The Mul-T-Lock Integrator is the next-generation development of the standard 7×7 system and provides a higher level of tamper-resistant security for a wide variety of locking applications. The Integrator consists of a unique high-precision seven-pin tumbler mechanism with a factory-ready patented key blank that includes a copy-protected oval cut (Figure 6.1). Special launcher pins are inserted into the relevant plug chambers of the cylinder that correspond to the oval key cut to complete the system's pin-tumbler security mechanism.

The special cuts on Integrator keys are available in three different configurations: internal, external, and twin cut. This enables a wide variety of keying options and a flexible master-keying hierarchy for different locking products and environments. The highly robust, tamperproof key and cylinder system can be keyed alike, keyed to differ, or master keyed according to preference.

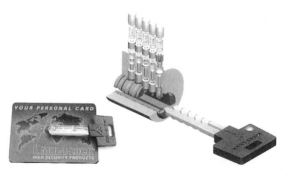

Figure 6.1 A cutaway view of a Mul-T-Lock and key.
(*Courtesy of Mul-T-Lock USA.*)

Additional key-combination cuts on the Integrator key blank can be made using the 7 × 7 key cutting machine and master system. The Mul-T-Lock Integrator is patent protected, thus ensuring that duplicate keys can only be manufactured by authorized Mul-T-Lock dealers, subject to presentation of the coded key registration card by the registered owner.

The Integrator system is ideal for environments that require a comprehensive, cost-effective locking solution without compromising on key-security standards for more sensitive areas. When combining the two systems together to secure a building or facility, the Integrator key and cylinder can be used as a first line of defense on the external perimeter of doors, and the 7 × 7 cylinders, which can be opened by Integrator keys, can be installed in internal doors and padlocks.

Schlage Primus

The Schlage Primus is one of Schlage Lock Company's newest high-security mechanical locking systems. It features a specially designed, patent-protected key that operates either the UL-listed high-security No. 20-500 Series cylinders or the controlled-access No. 20-700 Series cylinders. Both series are available in Schlage A, B, C/D, E, H, and L Series locks.

Both series are easily retrofitted into existing Schlage locks, and they can be keyed into the same master-key system and operated by a single Primus key. The Primus key is cut to operate all cylinders, whereas those that operate the standard cylinders won't enter a Primus keyway.

The Primus security cylinder is machined to accept a sidebar and a set of five finger pins, which, in combination with Schlage's conventional six-pin keying system, provides two independent locking principles operated by a Primus key. Hardened-steel pins are incorporated into the cylinder plug and housing to resist drilling attempts.

Primus security levels.   The Primus system features five different levels of security. Each level requires an appropriate ID card for key duplication.

Security level one. Primus cylinders and keys have a standard side milling allocated to level one and are on a local-stock basis. Level one also provides the flexibility of local keying into most existing Schlage key sections; it uses a Primus key to operate both systems. Key control is in the hands of the owner, who holds an ID card for the purpose of acquiring additional keys. Level one is serviced through qualified locksmiths who are trained at a Schlage Primus I center.

Security level one plus. Level one plus increases level one security by allocating a restricted side-bit milling to a limited number of local Primus centers, which also control keying and key records.

Security level two. At level two, service is carried out solely by original Schlage Primus distributors. Also, key duplication for the exclusive side-bit milling requires the authorized purchaser's signature. Factory keying is also available.

Security level three. At level three, the Schlage Lock Company maintains control of the inventory, keying, keying records, installation data, and owner's signature. Keys for geographically restricted side-bit millings can be duplicated only by the factory after the owner's signature has been verified.

Security level four. Level four provides all the restrictions of level three plus an exclusive side-bit milling.

Kaba

Kaba locks are dimple-key locks. There isn't just one but an entire family of different Kaba cylinder designs. Each one is designed to fill specific security requirements.

The family includes the Kaba 8, Kaba 14, Kaba 20, Kaba 20S, Saturn, Gemini, and Micro. Those that use numbers in their designations generally reflect not only the order of their development but also the number of possible tumbler locations in a cylinder of that particular design.

Handing and key reading

Handing.   The concept of *handing* is basic to an understanding of all Kaba cylinders. With Kaba cylinders, handing isn't a functional installation limitation, as you might expect. A left-hand cylinder will operate both clockwise and counterclockwise and function properly in a lock of any hand. The handing of Kaba cylinders refers only to the positions where the pin chambers are drilled.

The orientation of the two rows of side pins in a Kaba cylinder is staggered much like the disk tumblers in some foreign-car lock cylinders. There are two possible orientations of these staggered rows of side pins. Either row could start closer to the front of the cylinder. The opposite row then would start farther from the front. These two orientations are referred to as *right* or *left hand.* If the cylinder is viewed with the right side up, the hand of most Kaba designs is determined by the side whose row of pins begins farther from the front when viewing the face of the cylinder.

If a key is to do its job and operate a cylinder, obviously the cuts must be in the same positions as the pins in the cylinder. This means that there are two ways to drill dimples in the keys as well.

To determine the hand of a Kaba key, view it as though it were hanging on your key board. Notice that one row of cuts starts farther from the bow than the other. The row that starts farther from the bow determines the hand. If the row starting farther from the bow is on the left side, it is a left-hand key. If the row starting farther from the bow is on the right side, it is a right-hand key.

Occasionally, you may find a Kaba key with both right- and left-hand cuts. This is called *composite-bitted key*. It is used primarily in maison key systems (a type of master-key system).

Key reading. Determining the hand is the first step in key reading. Next, you need to know the order in which the various dimple positions are read.

The positions are always read bow to tip beginning with the hand side. After the hand side comes the nonhand side and then the edge. Remember to go back to the bow to start each row. For composite-bitted keys, use a Kaba key gauge to help mark the cuts of both hands.

All Kaba designs except the Micro have four depths on the sides; Micro has three. Kaba's depths are numbered opposite from the way one normally would expect. Number 1 is the deepest, and no. 4 is the shallowest. These depths can be read by eye with very little practice. The increment for side depths of the Kaba 8 and Kaba 14 is 0.0157 inch (0.4 millimeter). For the Gemini, it is 0.0138 inch (0.35 millimeter).

For the edges, reading is a bit different for the various Kaba designs. The Kaba key gauge is very useful for reading the edges. Find the section of the gauge that corresponds to the design of the Kaba key you're attempting to read (e.g., the Kaba 8). Place the key under the gauge so that it shows up through the edge slot. When the shoulder of the key hits the stop on the gauge, you're ready to read the positions of the cuts.

Because of the nonstandard cylinder drilling for the Kaba 8, the codes show two columns for the edge. There are only two depth possibilities on the edge of the Kaba 8 and the Kaba 14: cut = no. 2 and no cut = no. 4.

For the Kaba 14 and the new Kaba 8 using ME series codes, the edge is read differently because you know automatically which positions are involved in every case. These cylinders are all drilled with odd edges (positions 1, 3, 5, and 7).

Knowing the positions involved, the edge combination of these keys is notated in terms of depths rather than positions. A Kaba key gauge can be used to determine the positions of the no. 2 edge cuts. If there is a cut in position 5 only, the combination would be 4224. If there are cuts in positions 1, 3, and 7, the combination would be 2242. If there are cuts in positions 3 and 5, the combination would be 4224, etc. We know that there are four chambers in the cylinder, and they are in the odd-numbered positions. Therefore, we have to come up with four bittings in the edge combination. *Remember*: cut = no. 2 and no cut = no. 4.

For the Kaba Gemini, there are three active depths on the edge plus a high no. 4 cut used in master keying. Because there is no no cut on Gemini, a key gauge should not be necessary to read the edge combination. Right-hand stock keys always will have cuts in positions 3, 5, 7, and 9, and left-hand keys for factory master-key systems almost always will have cuts in positions 2, 4, 6, 8, and 10.

Bitting notation

Before the bitting comes an indication for the hand, "R" or "L." As mentioned earlier, all bittings are read and noted bow to tip. The key combination is broken up into separate parts corresponding to each row of pins in the cylinder. The first group of bittings is the hand side, the second group is the opposite side, and the third is the edge.

This holds true for all Kaba designs, but the guard-pin cut on a Saturn key (a no. 3 depth) is not part of the key combination and should be ignored for all phases of key reading.

Notation of composite bitted-key combinations isn't much different from that of regular keys. Composite-bitted keys have both right- and left-hand side cuts and often have both even and odd position edge cuts. Such keys normally are used only in selective key systems and maison key systems.

If most of the key system is made up of left-hand cylinders, that is the hand that is listed first. Conventionally, the edge bittings are all listed as part of the first line. The opposite hand bittings are written under those of the main hand.

Identifying nonoriginal keys

Because the dimensions of the key blanks for the Kaba 8, Kaba 14, and Saturn are identical, people in the field sometimes duplicate a key from one design onto another's key blank. This can lead to problems later, especially when quoting prices to a customer. If a key says "Kaba 8" but it really has been cut for a Kaba 14 cylinder, you might quote a Kaba 8 price and order a Kaba 8 cylinder only to find that you can't set it to the customer's key. To avoid confusion at a later date, a genuine key blank with a system designation such as "Kaba 8," "Kaba 14," or "Kaba Saturn" should only be used to make keys for that particular design. If you didn't sell the job originally, you always should verify the design by counting the dimples on the key.

If you need a new cylinder to match a nonoriginal key, or if the key was made poorly and you must make a code original to operate the lock properly, you also must be able to determine which Kaba design the key was cut for. This is done easily by counting the dimples on the sides, being careful not to overlook any of the tiny no. 4 cuts.

The Kaba 8 has two rows of four cuts (eight total). The Kaba 14 has two rows of five cuts (10 total). The Saturn has a row of three and a row of four (seven total). The Kaba Gemini uses a different key blank that is thicker and narrower. The dimples are oblong rather than round. It has a row of five cuts and a row of six cuts (11 total). Composite-bitted keys have exactly twice as many dimples on a side.

Medeco locks

Locks manufactured by Medeco Security Locks, Inc., are perhaps the most well-known high-security mechanical locks in North America. For this reason, much of this remaining section is devoted to reviewing how the various types of Medeco locks operate.

General operation.   Before studying in detail the specifications of Medeco locks, you must first understand how these locks operate. Medeco's 10 through 50 Series locks incorporate the basic principles of a standard pin-tumbler cylinder lock mechanism. A plug, rotating within a shell, turns a tailpiece or cam when pins of various lengths are aligned at a shear line by a key. Figure 6.2 shows an exploded view of a Medeco cylinder.

The plug rotation in a Medeco lock is blocked by the secondary locking action of a sidebar protruding into the shell. Pins in a Medeco lock have a slot along one side. The pins must be rotated so that this slot aligns with the legs of the sidebar. The tips of the bottom pins in a Medeco lock are chisel pointed, and they are rotated by the action of the tumbler spring seating them on the corresponding angle cuts on a Medeco key (Figure 6.3).

The pins must be elevated to the shear line and rotated to the correct angle simultaneously before the plug will turn within the shell. This dual-locking

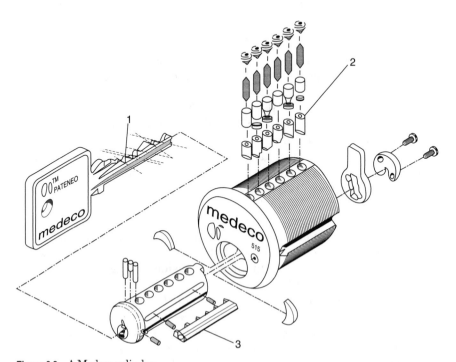

Figure 6.2   A Medeco cylinder.

Figure 6.3 A Medeco key.

principle and the cylinder's exacting tolerances account for Medeco's extreme pick resistance.

Medeco cylinders are also protected by hardened, drill-resistant inserts against other forms of physical attack. There are two hardened, crescent-shaped plates within the shell that protect the shear line and the sidebar from drilling attempts. There are also hardened rods within the face of the plug and a ball bearing in the front of the sidebar.

To fit within the smaller dimensions necessary in a cam lock, Medeco developed the principle of a *driveless rotating-pin tumbler*. It is used in the 60 through 65 Series locks. The tumbler pin and springs are completely contained within the plug diameter.

The plug rotation is blocked by the locking action of a sidebar protruding into the shell. The pins are chisel pointed and have a small hole drilled into the side of them. The pins must be rotated and elevated by corresponding angled cuts on the key so that each hole aligns with a leg of the sidebar and allows the plug to rotate. In addition, the cylinder is protected against other forms of attack by four hardened, drill-resistant rods within the face of the plug.

Despite the exacting tolerances and additional parts, Medeco cylinders are less susceptible to wear problems than are conventional cylinders. As in all standard pin-tumbler cylinder locks, the tips of the pins and the ridges formed by the adjacent cuts on the key wear from repeated key insertion and removal. In a Medeco lock, this wear has little effect on its operation because, in contrast to a standard lock cylinder, the tips of the pins in a Medeco lock never contact the flat bottoms of the key cuts. Instead, they rest on the sides of the key profile; thus the wear on the tips of the pins does not affect the cylinder's operation. Cycle tests in excess of 1 million operations have proven Medeco's superior wear resistance.

Medeco keys. There are four dimensional specifications for each cut on an original Medeco key. They are the cut profile, the cut spacing, the cut depth, and the cut angular rotation.

The profile of the cut on all original Medeco keys must maintain an 86-degree angle. This dimension is critical because the pins in a Medeco lock are chisel pointed and seat on the sides of the cut profile rather than at the bottom of the cut. Prior to June 1975, Medeco keys were manufactured with a perfect V-shaped profile. After that date, the keys were manufactured with a 0.015-inch-wide flat at the bottom.

Spacing of the cut on a Medeco key must be to manufacturer's specifications. For the 10 Series stock keys, Medeco parts KY-105600-0000 (old part 10-010-0000) and KY-106600-0000 (old part 10-011-0000), the distance from the upper and lower shoulder to the center of the first cut is 0.244 inch. Subsequent cuts are centered an additional 0.170 inch. For the 60 Series stock keys, Medeco parts KY-105400-60000 (old part 60-010-6000) (five pin) and KY-104400-6000 (old part 60-011-6000) (four-pin), the distance from the upper shoulder to the center of the first cut is 0.216 inch. Subsequent cuts are centered an additional 0.170 inch. The distance from the bottom shoulder to the center of the first cut is 0.244 inch on this and all Medeco keys. For the 60 Series thick-head keys, Medeco parts KY-114400-6000 (old part 60-611-6000) (five-pin) and KY-114400-6000 (old part 60-611-6000) (four-pin), the distance from the shoulder to the center of the first cut is 0.244 inch. Subsequent cuts are centered an additional 0.170 inch.

Standard Medeco keys in the 10 through 50 Series are cut to six levels with a full 0.030-inch increment in depth. Keys used in extensively master-keyed systems and on restricted Omega keyways are cut to 11 levels with a half-step 0.015-inch increment in depth. Because of size limitations, Medeco keys in the 60 through 65 Series are cut to four levels with a 0.030-inch increment in depth. Keys used in extensively master-keyed systems and Omega keyways are cut to seven levels with a 0.015-inch increment in depth.

In addition to these dimensions, each cut in a Medeco key may be cut with any one of three angular rotations. These rotations are designated as right (R), left (L), or center (C). When you look into the cuts of a Medeco key as illustrated, concentrating on the flats of the cut, you will notice that flats that are positioned perpendicular to the blade of the key are designated *center angles*. Flats that point upward to the right are designated *right angles*, and flats that point upward to the left are designated *left angles*. Right and left angles are cut on an axis 20 degrees from perpendicular to the blade of the key.

Keyways and key blanks.   The entire line of Medeco locks is available in numerous keyways. The use and distribution of key blanks of various keyways is part of Medeco's systematic approach to key control. Medeco offers four levels of security.

Signature program. Consumers can get new keys made simply by going to the locksmith from whom the lock was purchased. Keys for the Signature Program are restricted by Medeco to locksmiths who have contracted with the company.

Card program. A consumer is issued a card embossed with control data and that has space for a signature. The keys are owned by Medeco and are controlled by contract. An authorized locksmith will make duplicate keys only after verifying the card data and signature.

Contract restricted. A business or institution can enter into a contract with Medeco for a specially assigned keyway. The key blanks then can be ordered through any authorized Medeco distributor with appropriate authorization of the business or institution.

Factory program. The Factory program is Medeco's highest level of key control for consumers, businesses, and institutions. Keys are made from factory-restricted key blanks that are never sold by the factory. Duplicate keys are available only directly from the factory on receipt and verification of an authorized signature.

10 Series Pins.   Medeco bottom pins differ significantly from standard cylinder pins in four respects. The differences occur in the diameter, the chisel point, the locator tab, and the sidebar slot.

- Medeco pins have a diameter of 0.135 inch. This is 0.020 inch larger than the 0.115-inch-diameter pin in a standard pin-tumbler cylinder.

- All bottom pins are chisel pointed with an 85-degree angle. The tip is also blunted and beveled to allow smooth key insertion.

- The *locator tab* is a minute projection at the top end of the tumbler pin. The locator tab is confined in a broaching in the shell and the plug; it prevents the bottom pin from rotating a full 180 degrees. A 180-degree rotation causes a lockout because the sidebar leg is not able to enter the sidebar slot in the pin.

- The *sidebar slot* is a longitudinal groove milled in the side of the bottom pins to receive the sidebar leg. The slot is milled at one of three locations in relationship to the axis of the chisel point.

In 1986, the patent for the original Medeco lock system expired. Now other key-blank manufacturers can produce those key blanks. The original system, referred to by Medeco as "controlled," is still being used throughout the world. To allow greater key control, however, Medeco produced a new lock system called the *Medeco Biaxial* system and a newer one called *Medeco 3*.

Medeco's Biaxial lock system.   From the outside, Medeco Biaxial cylinders look similar to those of the original system. Internally, however, there are some significant differences.

Biaxial pins differ from original Medeco pins in three respects: the chisel point, the locator tab, and the pin length. Biaxial pins are made of CDA340 hard brass and are electroless nickel-plated. They have a diameter of 0.135 inch and are chisel pointed with an 85-degree angle. However, this chisel point is offset 0.031 inch in front of or to the rear of the true axis or center-line of the pin.

Fore pins are available in three angles: B, K, and Q. Aft pins are also available in three angles: D, M, and S. Pins B and D have a slot milled directly above their true centerline. Pins K and M have a slot milled 20 degrees to the left of the true centerline. Pins Q and S have slots milled 20 degrees to the right of the true centerline.

The locator tab, the minute projection limiting pin rotation, was moved to the side of the pin roughly 90 degrees. It is now located along the centerline of the pin opposite the area for the sidebar slot.

Medeco Biaxial key specifications.   There are four dimensional specifications for a Medeco Biaxial key: the cut profile, the cut depth, the cut spacing, and the cut angular rotation.

The cut profile of the Medeco Biaxial key remains at 86 degrees. However, keys in the 10 through 50 Series locks are cut to six levels with a full 0.025-inch increment in depth.

Spacing of the cut on Medeco Biaxial keys must be to manufacturer's specifications. Because Medeco Biaxial pins have the chisel point forward or aft of the pin centerline, the dimensional spacing on the Biaxial key blank can change from chamber to chamber.

From the shoulder to the center of the first cut using a fore pin (either B, K, or Q), the dimension will be 0.213 inch. From the shoulder to the center of the first cut using an aft pin (either a D, M, or S), the dimension will be 0.275 inch. Subsequent fore cuts and subsequent aft cuts are spaced at 0.170 inch.

While there are only three angular rotations on a key, each rotation can be used with either a fore pin or an aft pin. Angular cuts B and D are perpendicular to the blade of the key; B is a fore cut, and D is an aft cut. Angular cuts K and M have flats pointing upward to the left; K is a fore cut, and M is an aft cut. Angular cuts Q and S have flats pointing upward to the right; Q is a fore cut, and S is an aft cut. K, M, Q, and S angles are cut on an axis 20 degrees from perpendicular to the blade of the key.

7

Electronic Locks

Kaba 4000 Series

Kaba 4000 Series locks quickly replace any traditional key-operated door lock, providing code access and eliminating the need for keys, cards, and computers. Locks are easily programmed at the door. Each user or group of users is assigned a three- to six-digit code that provides secure access to all rooms they are authorized to enter (Figure 7.1). These locks are

- Easy to install—no wiring
- Easy to program—no computer required
- Ideal for establishing restricted areas within a facility
- Programmable at the lock with easy step-by-step instructions for access codes and set features
- Available to larger user bases because codes can be assigned to groups of users
- Equipped with a key override that accommodates major brands of interchangeable cores or key-in-lever cores
- Equipped with easily replaceable AA batteries that last up to 3 years
- Certified American National Standards Institute (ANSI) grade 1

The Simplex and E-Plex 5000 Series

The Simplex and E-Plex 5000 Series is the complete toolkit for all your access-control needs. Built on five decades of reliable and convenient keyless entry, this series provides everything from the simplicity of mechanical single-code access to the added security and features of electronic access control. Meet all your needs with the Simplex and E-Plex 5000 Series:

Figure 7.1 A Kaba 4000 Series lock.

- Cylindrical, mortise, or exit trim locking options
- PIN, PROX, or dual-credential access
- Single or multiple access codes
- Easy-to-use software available to simplify management of doors and access users
- Available features include audit trails, access schedules, remote unlock passage, and privacy
- Quick and easy through-bolt installation—no wiring to or through the door

Electronic access control

- No-wire installation—to or through the door
- Multiple authority levels give you control over who has access to specific features:
  - Master level—can perform all setup and programming functions
  - Manager/maintenance level—can administrate common programming functions
  - Authorized access user level
  - Service level—codes for single-event or single-day access
  - Visitor management—for access up to 1 year with expiry (E-Plex 5200 and 5700)
- Programmable lock functions
  - Entry lock—normally locked; passage mode can be activated by authorized operator
  - Storeroom lock—normally locked; passage mode deactivated
  - Residence lock—remains in passage until locked from interior with deadbolt or exterior key command
  - Privacy lock—activated from inside; restricts entry to privileged master- or manager-level user or key override
  - Remote unlock (optional)—from up to 100 feet away
- Easy-to-use software
  - Simplifies management of doors and access users
  - Match each door and user's access requirement to security needs
  - Point-and-click software lets you manage users

- Settings are easily cloned, permitting quick multilock setups
- Up to 32 holiday and vacation blocks (E-Plex 5200 and 5700)
- Up to 16 access schedules (E-Plex 5200 and 5700)
- Wireless communication
  - No cables are needed to upload and download your data
  - Communicates through infrared data-transfer function of a handheld personal digital assistant (PDA)
  - Fully contained E-Plex electronics module eliminates repair hassles

Four-bolt installation

- No wiring
- One-door prep
- Common template lets you change from mechanical to electronic or electronic to mechanical with ease.
- Universal exit trim mount—fits most leading brands of exit devices
- Installs easily on wood or metal doors
- Interchangeable knobs and levers
- Nonhanded/field reversible—fits both left- and right-hand doors
- Extra-heavy-duty locks provide unparalleled strength.
- ANSI/BHMA grade 1 certified
- Both housings are rugged and weather resistant—can be mounted outdoors; no installation gaskets required
- Outstanding technical support team with excellent reputation
- Three-year warranty

Installation

- Door thickness: Cylindrical: $1^3/_8$ to $2^1/_4$ inches (35–57 millimeters)
- Mortise: $1^3/_4$–$2^1/_4$ inches (44–57 millimeters)
- Exit trim: $1^3/_4$–$2^1/_4$ inches (44–57 millimeters)
- Preassembled to accommodate doors $1^5/_8$–2 inches (41–51 millimeters)
- Hardware kits included to accommodate thin and thick doors
- Nonhanded/field reversible
- Preassembled for left-hand door installations—easily changed in the field
- Minimum stile required: 5 inches (127 millimeters)

Finishes.   There are five attractive finishes to blend with any décor—bright brass (lifetime finish), satin brass (lifetime finish), satin chrome, black, and duranodic.

Key override options

- Key-in-lever/knob cylinders (cylinder included)
- Tailpieces included for compatibility with models by Abloy, Arrow, ASSA, Corbin Russwin, Kaba, Medeco, Sargent, Schlage, and Primus (see catalog for specific model details)
- Small-format interchangeable cores: Configured for best and compatibles (six- or seven-pin length)
- Large-format interchangeable cores: Available configurations for Schlage, Medeco, ASSA, Yale (six-pin), Sargent, and Corbin Russwin

Batteries (E-Plex models only)

- Four AA alkaline batteries
- Provides quick and easy battery replacement for institutions
- Lock data are retained if batteries fail
- Long battery life—up to 180,000 cycles

Certification and testing

- Accessibility standard: Americans with Disabilities Act (ADA)
- Latch fire rating: Three-hour UL/ULC fire door rating
- Durability: ANSI/BHMA A156.2 and .25 (where applicable)
- BHMA grade 1 certification (see BHMA Directory for listing of grade 1 certified Simplex and E-Plex products.)

Environmental parameters

- Indoor/outdoor approved
- Front housing: −31 to +151°F (−35 to 66°C)
- Rear housing: −31 to +130°F (−35 to 55°C)

Customizable user preferences (E-Plex models only)

- 2- to 20-second relock time
- Four- to eight-digit user codes
- Antitamper (disables lock after multiple invalid code entries)

Electromagnetic Locks

Because they're safe, strong, and practically maintenance free, electromagnetic locks are often part of an access-control system. They have no moving parts to wear out or to jam (Figure 8.1). And unlike deadbolts and other mechanical locks, these electrically controlled locks can be wired into fire alarms, smoke alarms, and sprinkler systems so that they'll unlock automatically in an emergency.

How They Work

Although electromagnetic locks can be actuated with a key, they operate on a different principle from mechanical locks. Instead of using a bolt or latch to secure a door, an electromagnetic lock relies on magnetism. A typical model consists of two basic parts: a rectangular electromagnet and a matching metal strike plate. The magnet, about the size of a brick, is installed on the door's header. The strike plate is installed on the door in alignment with the magnet.

When the door is closed and the magnet is powered (usually by a key or push-button code), the strike slaps against the magnet, and the door is held secure. Electromagnetic force is all that's needed to hold the door locked. Electromagnetic locks can provide more than three times the strength of a deadbolt—up to 3,000 pounds of holding force. In most cases, the door will give way before the lock does.

The lock's position at the top of the door is important. In a typical deadbolt installation, on the side of the door, the lock absorbs direct pressure during an attempted kick-in. An electromagnetic lock at the top of the door, however, receives little pressure when someone kicks near the center of the door. The door just flexes a little and absorbs much of the force.

Electromagnetic locks usually require 12 or 24 V dc power at 3 to 8 W. Some offer optional ac/dc operations at 12 or 24 V. The low-power requirements let you install enough standby battery power to allow for continued operation during a power failure.

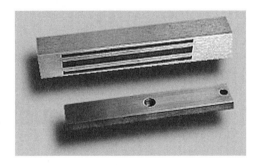

Figure 8.1 An electromagnetic lock. (*Courtesy of HighPower Security Products, LLC.*)

Thunderbolt Electromagnetic Locks

HighPower produces three models of electromagnetic locks and resells other types of electronic locks for special applications. For common metal-framed doors, the model of lock used depends on which way the door swings. If you are inside a room, the door swings out of the room, and you want the lock mounted inside the room, this is an *out-swinging* application. For out-swinging applications, use the Thunderbolt 1500 Electromagnetic Lock. If the door swings into the room and you want to the lock mounted inside the room, this is an *in-swinging* application. For in-swinging application, use the Thunderbolt 1560 Electromagnetic Lock. Both these locks have been tested to Underwriters Laboratories Standard 1034 for safety and can hold back 1500 pounds of force against the door. This force rating is excellent for holding back two or more people trying to push their way through the door.

Details about lock features

There is a circuit board in all Thunderbolt locks that can be ordered in different configurations. First, you have to decide what type of power you are going to run the lock on. The lock can be configured to work with either ac or dc power. For simple applications, such as using a transformer and a switch to control the lock, the ac version is useful. If you are using access-control circuits such as keypads or HighPower power supplies that have battery backup, order the dc version of the lock. The dc version is a much more common configuration because it works easier with most electronic controls. Although you have to order an ac or dc version of the lock from the factory, the voltages that the lock can run on are user-selectable. By changing a connection in the lock, Thunderbolt locks can run on either 12- or 24-V power.

In addition to the power requirements, you can order Thunderbolt locks with sensors that can be used to monitor door function and lock performance. A door position switch (DPS) may be ordered that senses if the door is open or closed. The DPS is a magnetic switch that is actuated when the door is closed. This switch is used often in control circuits in order to monitor the door from a remote location or to sound an alarm if the door is ajar for too long a time.

A magnetic bond sensor (MBS) also may be ordered. The MBS is used to tell if the lock is secure when the lock is powered. When the lock is powered (locked) and the armature is mated with the lock, this switch actuates. Using this switch, you can tell if the lock is energizing properly and can see if an obstruction has been put into the lock to prevent it from locking. Lastly, the lock can be ordered with a cover tamper switch (CTS). This switch is held closed by the cover of the lock. If the cover is pried off or tampered with, the switch opens and often is used to sound an alarm.

Powering the system

HighPower produces three different power supplies, but for a one-door system with simple controls, the Lightning 505 is a good choice. This is a 1-amp 12/24-V selectable power supply that has a built-in battery backup charging circuit. This power supply will switch over to battery power automatically when the main power goes out in order to keep the electromagnetic locks locked. This power supply also maintains the batteries, in that the batteries are charged continually during normal operation.

For more complex systems, HighPower uses the Lightning 4000 power supply. This power supply is a 4-amp unit that can be used to power multiple locks or locks with electronic controls. All the networked systems sold by HighPower come with this power supply. This supply also provides battery backup. HighPower customers who order a door controller with their system will have this power supply integrated with it in an enclosure along with the controller circuit board.

For low-cost systems, some customers opt to use a plug-in wall transformer with an ac-equipped lock. This type of power supply simply plugs into the wall and powers the lock directly. The drawback to this type of system is that the plug-in transformer often does not provide the amount of power regulation and voltage to make the lock run at maximum holding power. Also, this type of system does not allow for battery backup.

Controls for the locking system

A standard system usually has one control to unlock the door from the secured side and two methods of egress (exiting) from the nonsecured side. The control device is usually a proximity card reader; Wiegand-based keypad, bar-code, or magnetic-stripe reader; or a biometric device such as an iris, hand, or fingerprint reader.

HighPower often specifies the P-300 proximity card reader along with a "system controller." On the egress side of the door for systems with electromagnetic locks, a proximity sensor is used along with a Pushplate 100 PN switch. Using the proximity sensor, the system will unlock the door when someone walks up to the egress side of the door. On this side of the door also, the Pushplate 100 PN switch allows a user to press the switch if the proximity sensor fails or does not respond. This switch is time delayed and will provide a mechanical time delay that allows a person to exit even when the electronics fail. This is done for safety and to meet many local fire codes.

In addition, local fire codes may require that you tie the access-control system into the fire alarm system. In this case, when the fire alarm is tripped, the system cuts the power to the lock, allowing people to exit the building during a fire. HighPower usually sells a relay-board interface that can be used to connect the fire alarm to the fire alarm system. Please consult your local fire marshal and fire alarm panel manufacturer in any case where these locking systems are used.

A system controller is a circuit board that is used to perform certain locking functions and provide an interface between keypads, computers, door switches, and the lock. Currently, HighPower sells two controllers. One is for networked applications, and the second is for stand-alone applications.

The stand-alone controller, the HighPower 3000, can be used to run two doors. It provides all the basic functions required to connect two Wiegand devices (such as proximity readers, card readers, and biometrics), electromagnetic locks, electric strikes, and other security hardware, including video recorders. It is an excellent choice when you need a low-cost system for basic access and do not wish to log door entry. It also has many other uses and comes with an instruction manual that describes some interesting system applications.

These controllers usually are housed in an enclosure. The HighPower 3000 typically is packaged with a Lightning 4000 power supply and transformer. Batteries can be placed in both systems to provide a battery backup for each door. The system enclosure normally is mounted in a secure and inaccessible place such as a wiring closet or over a hung ceiling in a facility.

Hooking the system up

These systems should be installed by qualified installers and electricians. Normally, each system is sold with full documentation that shows the wire types and runs to use and how to connect each piece of the system to the system controller. HighPower attempts to make the electrical connections installable by anyone who can read a basic electrical schematic. In addition, HighPower analyses the hardware particulars in order make sure that you have any additional bracketing or mounting hardware that is required in your system. Along with the electronic documentation, the company provides installation procedures and installation templates for the hardware of your particular system.

Unusual applications

HighPower often tackles complex and unique systems along with its standard offerings. The company has produced specialized system controls and electronics for many customers at a reasonable cost and delivery. If you have special requirements, the company can help you to sort out the most efficient way to tackle your installation. Video cameras and controls can be integrated with any locking system.

Thunderbolt 1500

HighPower has created the Thunderbolt 1500, a slim-line 1,500-pound electromagnetic lock that is designed to provide the fastest installation times. Mounting the Thunderbolt is quick and easy. Extensive feedback from installers directed HighPower to incorporate an improved adjustable mounting system. This slotted mounting system allows freedom of movement to make adjustments during tough installations. A template is provided to quickly mark mounting holes for both the magnet and the armature.

Installers are discovering that the Thunderbolt 1500 is designed to maximize profits. The Thunderbolt 1500 has an epoxyless design that allows the magnetic coil to be unplugged from the unit and replaced should it become defective without having to uninstall the lock. This feature both reduces service time and provides improved value by keeping assembly costs low.

No more fooling around with wire splices. Connections are made to the Thunderbolt with screw-terminal blocks that speed up wire installation. With a single circuit board, the Thunderbolt can be configured quickly with a door position switch (DPS), a cover tamper switch (CTS), and a magnetic bond sensor (MBS). In addition, electronic spike and "kickback" surge suppression is standard with all models.

Installers love the Thunderbolt's single-piece cover. It slides into place, allowing a rapid and hassle-free installation. Having no exposed screws, the cover provides the highest level of tamper resistance. This modular cover allows installers to stock different color covers in order to quickly provide customers with the cosmetics specified.

HighPower is composed of people who care about your needs. HighPower employees have benefited from years of experience in the access control and security industry and are looking to constantly improve their products. HighPower backs all its products with hassle-free warranty service and is committed to making all your installations successful.

Thunderbolt 1560

The Thunderbolt 1560 is a 2-inch profile electromagnetic lock for in-swinging doors that combines unmatched ease of installation with solid product reliability. Designed to provide the fastest installation, the unit incorporates a unique tamper-resistant cover design and slotted mounting system with installation template. Providing 1500 pounds of holding force, all versions feature a replaceable dual-voltage coil that operates at both 12 and 24 V. Units can be equipped to operate using either ac or dc power and feature a surge and spike suppression circuit. All units include a fully assembled Z-bracket armature. The Thunderbolt 1560 is made in the United States and is backed with a 10-year manufacturer's warranty.

Intruder Alarms and Home Automation

Nowhere have recent advances in electronic and computer technology been more apparent than with security and home automation systems. Many types of systems that sell for under $1000 today weren't available 10 years ago at any price, and some of today's lowest priced systems are more effective and more reliable than ever.

To get your money's worth, however, you have to know what to look for. This chapter reviews a wide range of electronic security systems and devices. I'll explain why some of them can be useful and why many others can be costly nuisances. I'll also show you the basic installation procedure used for many types of alarms and home automation systems.

Burglar Alarms

More than 600 inmates of an Ohio prison were asked what single thing they would use to protect their homes from burglars. The most popular choice was a dog; the next was a burglar alarm. Other studies show that many police officers also believe a burglar alarm can make a home safer.

I favor installing intruder alarms, but they're not useful for everyone. To benefit from a burglar alarm, you and everyone in your home must learn how to operate it properly and must use it consistently. Everyone must remember to keep all windows and doors of the house closed when the system is armed. Many homeowners pay thousands of dollars for an alarm system only to discover that using it is too much trouble.

Contrary to popular belief, a burglar alarm doesn't stop or deter burglars. It only warns of their presence (if it's turned on during a break-in). Some burglar alarm sellers say that your having an alarm will make burglars think twice about trying to break into your home. Actually, it isn't your having the alarm that deters burglars; it's their belief that your home has an alarm that will stop them. Often, the only part of a burglar alarm that can be seen from outside is the window sticker. If you use alarm system window stickers, very few burglars will know whether you have or don't have a burglar alarm.

If you're like me and you have to have the real thing, you can either buy an alarm as a complete kit or get the components separately. The components are likely to include a control panel (Figure 9.1), a siren or bell, and various detection devices. In general, complete kits are less expensive than separate components, but by mixing and matching components, you can create a more effective burglar alarm system.

Detection devices (or *sensors*) are the eyes and ears of a burglar alarm system. They sense the presence of an intruder and relay the information to the system's control panel, which activates the siren or bell. Today, you have more detection devices to choose from than ever before, but if you use the wrong type the wrong way, you'll have a lot of false alarms.

Some detection devices respond to movement, some to sound, and others to body heat. The principle behind each type is similar: When an alarm system is turned on, the devices sense and monitor a "normal" condition; when someone enters a protected area, the devices sense a disturbance in the normal condition and trigger an alarm.

Most detection devices fall within two broad categories: perimeter and interior. Perimeter devices are designed to protect a door, window, or wall. They detect an intruder before entry into a room or building. The three most common perimeter devices are foil, magnetic switches, and audio discriminators. Interior (or space) devices detect an intruder on entry into a room or building. The five

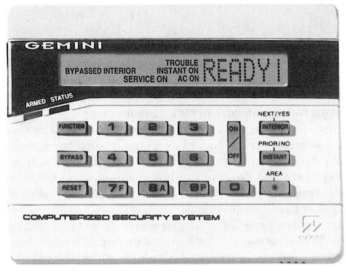

Figure 9.1   An alarm system control panel. (*Courtesy of NAPCO Security Systems.*)

most common interior devices are ultrasonic, microwave, passive infrared, quad, and dual-tech detectors.

Foil

You've probably seen foil on storefront windows. It's a thin, metallic, lead-based tape, usually $1/2$ to 1 inch wide, that's applied in continuous runs to glass windows and doors. Sometimes foil is used on walls. Like wire, foil acts as an electrical conductor to make a complete circuit in an alarm system. When the window (or wall or door) breaks, the fragile foil breaks, creating an incomplete circuit and triggering the alarm.

Usually foil comes in long adhesive-backed strips and is applied along the perimeter of a sheet of glass or dry wall. Each end of a run must be connected to the alarm system with connector blocks and wire. Foil is popular in stores because it costs only a few cents per foot, and acts as a visual deterrent.

There are three major drawbacks to foil:

1. It can be tricky to install properly.

2. It breaks easily when a window is being washed.

3. Many people consider it unsightly.

Whether or not you like foil, foil alone is rarely enough to protect a home. Other detection devices also should be used.

Magnetic switches

The most popular type of perimeter device is the magnetic switch. It's used to protect doors and windows that open. Magnetic switches are reliable, inexpensive, and easy to install.

As its name implies, the device consists of two small parts: a magnet and a switch. Each part is housed in a matching plastic case. The switch contains two electrical contacts and a metal spring-loaded bar that moves across the contacts when magnetic force is applied. When magnetic force is removed, the bar lifts off one of the contacts, creating an open circuit and triggering an alarm condition.

In a typical installation, the magnet is mounted on a door or window, and the switch is aligned about $1/2$ inch away on the frame. When an intruder pushes the door or window open, the magnet is moved out of alignment. Some magnetic switches are rectangular, for surface mounting. Others are cylindrical, for recessed mounting in a small hole. The recess-mounted types look nicer because they're less conspicuous, but they're a little harder to install.

One problem with some magnetic switches is that an intruder can defeat them by using a strong magnet outside a door or window to keep the contacts closed. Some models can be defeated by placing a wire across the terminal screws of the switch, jumping the contacts. Another problem is that if a door is loose fitting, the switch and magnet can move far enough apart to cause false alarms.

Wide-gap reed switches can be used to solve those problems. Because reed switches use a small reed instead of a metal bar, they're less vulnerable to being manipulated by external magnets. The wide-gap feature allows a switch to work properly even if the switch and magnet move from 1 to 4 inches apart. Some magnetic switches come with protective plastic covers over their terminal screws. The covers thwart attempts at jumping. Most types of magnetic switches cost just a few dollars each.

Audio discriminators

Audio discriminators trigger alarms when they sense the sound of glass breaking. The devices are very effective and easy to install. According to a survey by *Security Dealer* magazine, over 50 percent of professional alarm installers favor audio discriminators over all other forms of glass break-in protection.

By strategically placing audio discriminators in a protected area, you can protect several large windows at once. Some models can be mounted on a wall up to 50 feet away from the protected windows. Other models, equipped with an omnidirectional pickup pattern, can monitor sounds from all directions and are designed to be mounted on a ceiling for maximum coverage.

A problem with many audio discriminators is that they confuse certain high-pitched sounds—such as keys jingling—with the sound of breaking of glass and produce false alarms. Better models require both the sound of breaking glass and shock vibrations simultaneously to trigger their alarm. This feature greatly reduces false alarms.

Another problem with audio discriminators is that their alarm is triggered only if glass is broken. An intruder can bypass the device by cutting a hole through the glass or by forcing the window sash open. Audio discriminators work best when used in combination with magnetic switches.

Ultrasonic detectors

Ultrasonic detectors transmit high-frequency sound waves to sense movement within a protected area. The sound waves, usually at a frequency of over 30,000 hertz, are inaudible to humans but can be annoying to dogs. Some models consist of a transmitter that is separate from the receiver; others combine the two in one housing.

In either type, the sound waves are bounced off the walls, floor, and furniture in a room until the frequency is stabilized. Thereafter, the movement of an intruder will cause a change in the waves and trigger the alarm.

A drawback to ultrasonic detectors is that they don't work well in rooms with wall-to-wall carpeting and heavy draperies because soft materials absorb sound. Another drawback is that ultrasonic detectors do a poor job of sensing fast or slow movements and movements behind objects. An intruder can defeat a detector by moving slowly and hiding behind furniture. Ultrasonic detectors are prone to false alarms caused by noises such as a ringing telephone or jingling keys. Although they were very popular a few years ago, ultrasonic detectors are a poor choice for most homes today. They can cost over $60; other types of interior devices cost less and are more effective.

Microwave detectors

Microwave detectors work like ultrasonic detectors, but they send high-frequency radio waves instead of sound waves. Unlike ultrasonic waves, these microwaves can go through walls and be shaped to protect areas of various configurations. Microwave detectors are easy to conceal because they can be placed behind solid objects. They are not susceptible to loud noises or air movement when adjusted properly.

The big drawback to microwave detectors is that their sensitivity makes them hard to adjust properly. Because the waves penetrate walls, a passing car can prompt a false alarm. Their alarms also can be triggered by fluorescent lights or radio transmissions. Microwave detectors are rarely useful for homes.

Passive infrared detectors

Passive infrared (PIR) detectors became popular in the 1980s. Today, they are the most cost-effective type of interior device for homes. A PIR detector senses rapid changes in temperature within a protected area by monitoring infrared radiation (energy in the form of heat). A PIR detector uses less power, is smaller, and is more reliable than either an ultrasonic or a microwave detector.

The PIR detector is effective because all living things give off infrared energy. If an intruder enters a protected area, the device senses a rapid change in heat. When installed and adjusted properly, the detector ignores all gradual fluctuations of temperature caused by sunlight, heating systems, and air conditioners.

A typical PIR detector can monitor an area measuring about 20 by 30 feet or a narrow hallway about 50 feet long. It doesn't penetrate walls or other objects, so a PIR detector is easier to adjust than a microwave detector, and it doesn't respond to radio waves, sharp sounds, or sudden vibrations.

The biggest drawback to PIR detectors is that they don't "see" an entire room. They have detection patterns made up of "fingers of protection." The spaces outside and between the "fingers" aren't protected by the PIR detector. How much of an area is monitored depends on the number, length, and direction of zones created by a PIR detector's lens and on how the device is positioned. Figure 9.2 illustrates various detection patterns.

Many models have interchangeable lenses that offer a wide range of detection pattern choices. Some patterns, called *pet alleys*, are several feet above the floor to allow pets to move about freely without triggering the alarm. Which detection pattern is best for you will depend on where and how your PIR detector is being used.

A useful feature of the latest PIR detectors is *signal processing* (also called *event verification*). This high-tech circuitry can reduce false alarms by distinguishing between large and small differences in infrared energy.

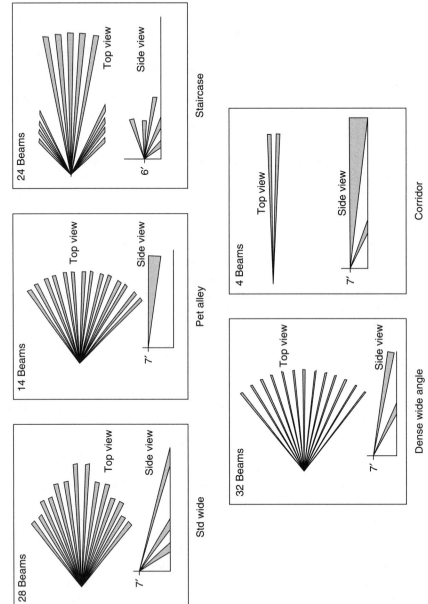

Figure 9.2 PIR detectors have various detection patterns.

Quads

A quad PIR detector (or quad, for short) consists of two dual-element sensors in one housing. Each sensor has its own processing circuitry, so the device is basically two PIR detectors in one. A quad reduces false alarms because, to trigger an alarm, both PIR detectors must detect an intrusion simultaneously. This feature prevents the alarm from activating in response to insects or mice. A mouse, for example, may be detected by the "fingers of protection" of one of the PIR detectors but would be too small to be detected by both at the same time.

Dual techs

Detection devices that incorporate two different types of sensor technology into one housing are called *dual-technology devices* (or *dual techs*). A dual tech triggers an alarm only when both technologies sense an intrusion. Dual techs are available for commercial and residential use, but because they can cost several hundred dollars, dual techs are used more often by businesses. The most effective dual tech for homes is one that combines PIR detectors and microwave technology.

For this type of dual tech to trigger an alarm, a condition must exist that simultaneously triggers both technologies. The presence of infrared energy alone or movement alone would not trigger an alarm. Movement outside a wall, which ordinarily might trigger a microwave, for example, won't trigger a dual tech because the PIR element wouldn't simultaneously sense infrared energy.

Professional versus Do-It-Yourself Alarms

Until recently, there were sharp differences between professional and do-it-yourself home alarms. A professional system was more reliable and harder for burglars to defeat. Do-it-yourself models were cheaper and easier to install but weren't very reliable. They often produced false alarms and failed to detect intruders.

Many of today's do-it-yourself systems, however, not only are simple to install, but they are also very effective and reliable. If you know what to look for, you may find that a do-it-yourself model is a better buy for you than a comparable professionally installed model.

Few installers buy home alarms in large enough volume to compete with the prices of hardware, department, and electronic products chain stores. To make it harder for consumers to compare an alarm installer's prices with those found in chain stores, many installers sell lesser known (but more expensive) brands. This helps installers to promote a sales pitch claiming that although their systems cost more, they're better.

The *hardwired* alarm system always has been the alarm of choice among professional installers. This system requires wire to be run from its control panel to each of its sensors (and to its siren or bell). Hardwiring will give you maximum reliability and will allow you to find and stop false alarms quickly.

However, running wires in a way that is both aesthetically pleasing and hard for burglars to defeat can be tricky and time consuming. For this reason, few hardwired models are made for do-it-yourselfers.

Wireless alarms are easy to install and therefore very popular among do-it-yourselfers. They rely on radio waves—instead of wire—for communication between the control panel and the sensors. Some wireless alarms can be installed with nothing more than a screwdriver. (With some "wireless" models, you still have to run wire from the siren to the control panel.)

Until recently, the biggest problem with do-it-yourself wireless alarms was that they weren't "supervised"; only expensive high-tech models included supervising circuitry. A *supervised* alarm is one that regularly checks its sensors to confirm that they're communicating properly with the control panel. With this capability, you know immediately if a sensor is broken, if the alarm's battery is low, or if a protected door or window has been left open.

A self-contained alarm is a single unit that is usually shaped like a VCR; its siren, motion detector, and other sensors are all built in. Installation is just a matter of positioning it on a sturdy table or shelf in a way that allows the unit to protect a selected area and then plugging it into an electrical outlet. The unit sounds an alarm when someone enters the protected area.

Two advantages of a self-contained unit compared with other types of home alarms are ease of installation and portability. You can take the unit with you when you move or when you're traveling. Its main disadvantage is that it protects only one area at a time.

Some self-contained units can be wired to use door and window sensors, glass-break sensors, an external horn, and other accessories. These features basically allow you to turn the self-contained unit into a hardwired system, which gives you tremendous flexibility.

What You Can Do That the Pros Won't

A seasoned professional can install an alarm system creatively so that it can thwart even the most sophisticated burglars. However, creativity doesn't come cheap. Unless you're willing to pay hundreds of dollars on your home security, a professional installer isn't likely to do much more than you may be able to do for yourself. In a typical home installation, the installer will place a sensor on each entry door and a motion detector near the main entry door.

Because many burglars know its shortcomings, a typical professional installation actually may *increase* your risk of being burglarized. When sophisticated burglars see certain alarm system stickers on homes, they know many ways they can get in, move about inside the homes, and get out undetected. By installing your own system, not only can you save money on installation, but you also can make sure that all entry points are protected adequately. (*Tip*: Never use the window stickers that come with your alarm; they give away too much information about your system. Buy no-name window stickers from a home-improvement center or hardware store.)

Most manufacturers of do-it-yourself alarms try to make it easy for you to custom install their systems by providing heavily illustrated installation manuals, free installation videos, and technical advice by telephone. Some companies have toll-free numbers solely for providing installation and operating guidance.

Tricks of Hardwiring

Although hardwired systems generally are more reliable and less expensive than their wireless counterparts, few laypersons like to install hardwired alarms. Sometimes, getting a length of wire from a control panel to the sensors can be tricky. Here are some tips that might help:

- When running wire from one floor to another, try using the existing openings used by plumbing or vents.

- If you have to drill a hole to get wire from one floor to another, consider drilling in a closet or another place that won't be noticeable. As a last resort, consider drilling as close to a corner as possible.

- Try running wire above drop ceilings.

- Try running wire under wall-to-wall carpet as close to the walls as possible (not in high-traffic pathways).

- If you can't hide the wire you're running, consider running it through plastic strips of conduit. (Conduit not only makes the run look neater but also protects the wire.)

- If you can't hide the wire and aren't using conduit, try to run the wire close to the baseboard.

- When running wire without conduit, you may need to staple the wire. Use rounded staples only; flat-back staples may cut into the wire and cause problems.

Home Automation Systems

Although locks, light, sound, and other elements play a part in home security and safety, each of these elements must be controlled separately in most homes. By using a home automation system, however, you can make several or all of the systems and devices in your home work automatically to provide more security, safety, and convenience.

Home automation is a generic term that refers to any automated technology used in homes—a sprinkler system that shuts itself off during a rain storm, automatic lights that come on when someone pulls into your driveway, and so on. If the right attachments are used, all home automation systems can perform many of the same functions. However, there are important differences among the three basic types of systems:

1. Programmable controller

2. Smart House integrated system

3. X-10–compatible modules

Programmable controller

A more versatile type of home automation system is one that uses a programmable controller that is integrated into your home's electrical power line. This type of system allows all your automation devices to work together under central control. By touching a keypad in your bedroom, for instance, you could turn down the heat in your home, arm your burglar alarm, and turn on your outdoor lights, or you could use your programmable controller to make all those things occur automatically every night at a certain time. However, it can cost up to $20,000 to have a full-blown power-line system installed in a home.

Smart House integrated system

One of the latest and most sophisticated home automation systems is the Smart House. Although the term *smart house* sometimes is used to refer to a wide variety or a combination of home automation systems, it's actually a brand name for a unique system of automating a home. The Smart House integrates a unique wiring system and computer-chip language to allow all the televisions, telephones, heating systems, security systems, and appliances in a home to communicate with each other and work together.

If your refrigerator door has been left open, for instance, the Smart House could signal your television set to show a picture of a refrigerator in the corner of the screen until you close the door. A smoke detector in the Smart House could signal your heating system to shut down during a fire. Its communication ability is one of the most important differences between the Smart House and all other home automation systems. The basic installation cost of a Smart House system is about the same as that of a power-line system, but with a Smart House system you also may need to purchase special appliances.

How the Smart House system works.   To understand how Smart House technology works, it's important to realize that the technology was the result of a joint effort among many appliance manufacturers, security system manufacturers, and home building and electronics trade associations. All of them agreed on standards that allow special appliances and devices to work in any Smart House.

A Smart House uses a system controller instead of a fuse panel and Smart blocks instead of standard electrical outlets. Appliances that are designed to work in a Smart House are called *Smart appliances*; all of them can be plugged into any Smart block. The same Smart block into which you plug your television, for instance, can be used for your telephone or coffee pot. When a Smart appliance is plugged into a Smart block, the system controller receives a code to release power, and it coordinates communication between that appliance and the other Smart appliances.

A big difference between standard electrical outlets and Smart blocks is that electricity is always present in the standard outlets. If you were to stick a metal pin into one of your standard outlets, you would get an electric shock. If you were to stick a pin into a Smart block, you wouldn't get shocked because no electricity would be present. Only a device that has a special computer-chip code can signal the Smart House system controller to release electricity to a particular Smart block—unless you override the signal.

With a Smart House, you have the option of programming any or all of the Smart blocks to override their need for a code. Such an option allows you to use standard appliances in your Smart blocks in much the same way that you use your electrical outlets now. Standard appliances can't communicate with each other or with Smart appliances. You might want to override a Smart block if some of your appliances aren't Smart appliances.

Because the Smart House is a new technology, very few Smart appliances are available. If the technology becomes more widely used, the demand for Smart appliances will increase.

Home automation controllers

With either a Smart House or a power-line system, you need only one controller to make the system do anything you want it to do. For convenience, however, you might want controllers installed at several locations in your home. In addition to a keypad, you can use your telephone, a computer, or a touch screen for remote control of your system. A touch screen looks like a large television that is mounted into a wall. It displays a "menu" of your options—lighting, security, audio, video, temperature controls, and so on—and you can make your selection just by touching the screen. If you were to touch "Security," for example, a blow-up of the floor plan of your home may appear on the screen, and you would be able to see whether any windows or doors are open, whether your alarm system is on or off, and other conditions related to your home's security. You also would be able to secure various areas of your home just by touching the screen.

X-10–compatible home automation systems

You may think that a comprehensive electronic security system costs thousands of dollars and requires a professional to install it. Some security systems do, but companies such as IBM, Leviton, RCA, Heath, Radio Shack, Sears, and Stanley offer effective, low-cost home security products that are X-10–compatible. X-10–compatible devices are easy to for security professionals and do-it-yourselfers to install.

They all share the same X-10 technology—a system that enables security and home automation components to operate using house wiring and compatible radio frequencies. This means that you can mix components from several manufacturers. Best of all, you can create an effective X-10 automation and security system for less than $200 and expand it later by selecting from a wide variety of components.

With an automation and security system in place, you can operate house lights and appliances or trigger the alarm siren with a hand-held remote control device. You can even adjust your thermostat, turn on the coffee pot, and listen for intruders—all from a phone booth across town.

How it works.   X-10 products require little or no wiring. In most cases, you simply plug devices into existing wall outlets or screw them into light sockets and then turn a couple of dials. A typical X-10 system includes a variety of controllers, modules, and switches. Each device transmits or receives high-frequency signals, which travel along your home's electrical wiring or through the air as radio waves. Modules receive these signals from controllers to operate lights, alarms, and appliances.

Most controllers, modules, and switches must be programmed with house and unit codes. Each component's face features a red dial labeled A through P and a black dial labeled 1 through 16. The red dial sets the house code, which identifies the devices as part of the same system and prevents accidental operation by a neighbor with an X-10–compatible setup. The black dial controls the unit code, which makes appliances work together or on their own. Set a group of lights to the same unit code, and they'll switch on or off simultaneously. X-10–compatible systems provide 256 possible house/unit-code combinations.

A basic X-10 installation.   First, go through each room in the house and decide which doors, windows, and areas you want protected from intruders. Also decide which lights and appliances you want the system to control. Once you install a basic system, you can add components later as your security needs grow. After choosing components, use a screwdriver to set house and unit codes.

The plug 'n power supervised security console dialer with hand console controls the system. The dialer operates up to 16 groups of lights, appliances, and alarm sensors. When the alarm is tripped, the dialer also phones up to four numbers and plays a recorded message. The pocket-sized hand console controls up to four lights and appliances.

Install the dialer close to an electrical outlet and phone jack but beyond an intruder's easy reach—a nightstand in the master bedroom usually works best. After flipping the dialer's mode switch to "Install," attach a 9-V battery as a backup in case of power failure. Plug the unit in, raise its antenna, and push the unit's earphone into it's jack. Then run the phone cord from the unit to the phone jack.

Next, program four emergency numbers into the dialer the same way you store speed-dial numbers on a telephone. Finish off by recording a 13-second message, such as, "A burglary may be in progress at John Smith's house. The burglar alarm was tripped. Press zero to listen in, and call the police if necessary."

Before testing the alarm, call the first person on the programmed recording list and explain the system's operation. Set the dialer's mode switch to "Run 2," and press "Arm" on the hand-held remote to arm the system. Then press the remote's "Panic" button to trip the alarm. Be sure that you can hear the dialer's built-in siren throughout the house or office.

Once the alarm goes off, the dialer contacts the first number. After the listener hears your message, he or she can press zero to shut off the alarm and listen for sounds of an intruder. When you finish the test, press "Disarm" on the remote, and flip the dialer's switch back to "Install."

Alarm without wires

Designed for mounting on doors and windows, wireless alarm transmitters cost just a few dollars each. Before installing them, open the battery compartment and slide the switch to "NC" (normally closed. This programs the unit to trigger the alarm when the door opens, whereas selecting "NO" (for normally open) sounds the alarm when the door closes. Then insert a 9-V battery.

Mount the alarm transmitter with Velcro fastening tape, and use screws to attach the switch to the door and magnet to the door frame. Be sure the switch and magnet align perfectly. The transmitter sounds the alarm when the electrical connection between them breaks, so an imprecise installation could cause false alarms.

Wall outlets and modules

Wall outlets control lamps or appliances plugged into them. Wall switches operate indoor and outdoor overhead lights with an X-10 controller. Be sure to turn off your home's electricity at the main circuit breaker before installing outlets and switches.

Remove the existing wall outlet's cover plate, and pull the outlet out of the electrical box. Then disconnect the wires running from the box to the outlet. When connecting the new outlet, simply match the wire colors. If no ground wire extends from the electrical box, connect the module's green wire to the box. Push the module into the box, and install the new cover plate. Then follow the same steps to install a remote switch.

You can install the Anywhere Wall Switch on any flat surface in the house. It requires four AAA batteries and operates four light fixtures without running wires. Use Velcro fastening tape or screws to mount the module. To operate lamps, a coffee maker or other small appliance, plug the Two-Prong Polarized Lamp or Appliance Switch into a wall outlet. You can control any appliance through the system once it's plugged into the switch.

Troubleshooting

If your X-10-based system works poorly even after proper installation, there are two likely causes: lack of phase coupling and power-line noise, also called *interference*. Power enters your home as 220/240-V service from two hot wires, called *phase A* and *phase B*, and your outlets are divided between them. If you plug a transmitter into a phase A outlet and a receiver into a phase B outlet, the transmitter may have to send its signal to the outdoor transformer before it reaches the receiver. By the time the signal arrives, it may be too weak for the unit to work properly. Check the circuit box diagram to find out each outlet's phase, and make sure that controllers and switches match. An electrician also can install a phase coupler at the circuit breaker to bridge the phases.

Fluorescent lights, computers, televisions, and other appliances can produce electrical noise that interferes with X-10 signals. If your system has a problem, unplug appliances one at a time to find the culprit. Eliminate the noise by keeping the offending appliance unplugged or installing noise filters. If fluorescent lights are the only source of line noise, try replacing their ballasts because some versions produce less noise.

10

Access-Control Systems

As the name implies, an *access-control system* controls the flow of pedestrian and vehicular traffic through entrances and exits of protected areas or premises. Basically it's an electronic way of controlling who can go where and when, sometimes keeping records of when a protected area has been entered. Typically, such a system uses coded cards, biometric readers, or magnetic keys.

Card systems are used frequently at hotels for entering rooms. The coded information on the key cards works only to the extent that the person controlling the access-control system wants it to. He or she can make the card invalid at any time, which is useful in a hotel if a guest doesn't pay the bill on time.

Biometric systems control access by comparing some physical aspect of a person with information on file. Hand and finger scanning technologies are popular among biometric systems. See figure 10.1. There's more on biometric systems in this chapter.

A popular type of magnetic key is the Corby data chip. It's a dime-shaped sealed steel canister that contains sophisticated electronics to store a personal identification number. Its size allows it be attached easily to any smooth surface, including photo ID cards, badges, and keychains. The design also protects the electronic circuits inside the canister from dirt, moisture, corrosion, and static discharge.

Touching a data chip to the reader instantly transfers a 46-bit stream of digital data that gives the user access to the secured area. Unlike keys or other security cards, the data chip is user forgiving—it doesn't need to be aligned precisely to transfer its digital data. See figure 10.2.

Along with the chip reader, you'll need a lock of some type. Usually, it's an electric strike or magnetic lock. See figure 10.3. Often an intercom or telephone system is part of the system too.

Figure 10.1 A handreader is a popular biometric device.

Figure 10.2 In the Corby system doors are unlocked by placing a dime-shaped key over the reader.

Figure 10.3 Parts of a typical electromagnetic lock.

Access Control for an Apartment

Access-control systems can be designed in many different ways depending on your creativity and the needs of your customer. Here's one configuration used for an apartment building. When someone comes to one of the main doors during the daytime, the person can enter the building by pushing a push pad (or just by pulling the door open) because the door is unlocked. See figure 10.4. This gives a visitor access to the manager's office, but not to the apartments. There's a separate door to the apartments, and it's locked 24 hours per day. At night, the main doors are also locked, and only people with keys may enter, such as tenants and people to whom tenants have given keys. After entering the lobby, a guest uses the in-house phone to call a tenant and ask to be buzzed in. See figure 10.5. Because a video camera watches the lobby 24 hours a day, the tenant can turn to a certain television channel to see who's calling (Fig. 10.6). A tenant who doesn't want to let a caller in can deny entry and call for help. See figure 10.7.

Introduction to Biometrics (by eKeyUSA Systems)

Biometrics is concerned with the measurement and identification of individual traits and characteristics. At the moment, unique attributes, such as fingerprints, voice, face and eye recognition can be measured and are subsequently used for the purpose of identification or verification. Different techniques are currently available: finger scanners, iris recognition, face recognition, voice verification, to name just a few. Usually, individual needs determine the most suitable method.

To put it into a nutshell, biometric identification always checks who you are, instead of what you know (passwords, codes, PIN) or what you possess (key, card). Nowadays, when people are regularly flooded passwords, biometric identification offers a convenient and attractive alternative in order to check someone's identity.

Figure 10.4 A push plate makes it easy to open unlocked doors.

Figure 10.5 Some access-control systems require guests to call someone to be buzzed into the living (or other protected) areas.

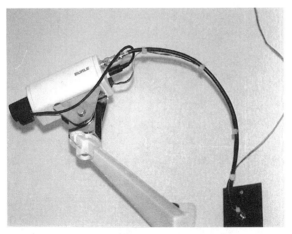

Figure 10.6 A visible closed-circuit television camera is useful in access control systems; it lets unauthorized persons know they're being watched.

Figure 10.7 A television in an apartment can be used to monitor certain areas.

What is biometric authentication?

Biometric authentication is the automatic identity verification or identification, based on individual physiological or behavioral characteristics. The authentication process is done via computer technology without violating anyone's privacy. Biometric characteristics are recoded first and then codified into a set of data which will be finally compared to the template. Nowadays, the offered biometric solutions include finger scans, face or iris recognition, and many more.

What is applied biometric?

If biometric is used for security matters, such as access control for example, it is possible to check human being's identity electronically. Therefore not only less security personnel is necessary, but also the average access time as well as the probability of potential misuse is reduced. Biometric solutions ensure an extremely high standard for different security applications.

Where is biometric used nowadays?

At the moment biometric applications can be found in many different areas, such as during border control, to secure computer networks and financial transactions, as well as access control or time recording. Available products range from door locks, safes, PC system (including safeguarding files and confidential content), computer mice, keyboards, web cameras, and time and attendance check. Biometric systems provide excellent convenience and a very high level of security.

How come that passwords and PINs decrease the level of security?

In order to consider a password as being "safe," it should be as long as possible, not to be found in a dictionary and even include a few special characters such as §, +, ?. Needless to say, that you are not allowed to write it down and in addition, it should be changed every 3 months. Since there are passwords for many areas (mobile phones, computer, online banking, debit card, etc.), users tend to use always the same all-round password. Generally speaking, they are rather machine friendly, instead of user friendly.

Why use biometric technology?

The answer to this question is simple! Individual biometric characteristics are unique and cannot be transferred. Security applications based on biometric identification/verification offer a high level of convenience. Furthermore information cannot be lost, stolen or forgotten. Finally, it is possible to reduce administration and maintenance costs considerably.

Which advantages does biometric recognition offer, as opposed to traditional security systems and methods?

- Increased convenience
- Eliminates the inconvenience of remembering passwords
- Eliminates the need for password maintenance
- Eliminates shared passwords
- Reduces maintenance and security costs
- Eliminates the use of even more keys
- Eliminates the use of remembering your swipe card
- Increased security for confidential files
- Deters fraudulent account access
- Biometrics cannot be lost or forgotten
- Extremely difficult to forge
- Highly accurate
- Uses distinctive features
- Faster Login

How do biometric devices work?

Biometric devices compare a recorded and encrypted set of data with a template.

How does biometrics protect personal information from unauthorized disclosure?

A biometric record is an encrypted template stored in digital form. The record itself cannot be used to reveal someone's identity; the probability of fraud is reduced to a minimum. With regard to the level of security, biometric offers an extremely high level of protection.

Is there a possible health risk?

No! The methods used to gather the personal biometric data are noninvasive and pose absolutely no threat to physical health or well-being. When properly installed biometric devices enhance both security and ease of access for the individual user. Since the eighties, when biometric systems were used the first time, definitely no health hazards have been found.

Are there any predictions for the future of biometrics and fingerprint recognition?

JP Morgan has predicted that the size of the biometric authentication market is going to increase steadily and will become standard for authentication and systems access. At the moment, fingerprint recognition is still in its early adoption phase but is gaining in awareness and acceptance rapidly. Passwords, PINs, cards, and signatures are eventually going to be replaced entirely.

Which measuring sizes can be used for verification of the performance potential of a biometric authentication system?

False acceptance rate (FAR). FAR represents the number of times when unauthorized persons are accepted as authorized. As the false acceptance can lead to damages or fraud, the FAR is an extremely important rate for security.

Failure to recognize rate (FRR). The FRR is the number of times an authorised person is rejected. The FRR can be used as a rate for convenience, as the failure to recognise normally is very troublesome.

Failure to enroll (FTE). FTE indicates the number of persons which, on average, can not be enrolled at all. If someone already enrolled is rejected after several trials, one speaks about a *failure to acquire* (FTA). These FTA rates can be added to the FRR and therefore do not have to be listed separately.

Why are not all biometric systems equally secure?

Generally speaking, the degree of security of a system not only depends on the technology and the algorithms, but also on the overall security concept and the parameters set. The potential solution for an individual customer has to be adjusted to the respective system. However, any adjustments do influence the FAR and FRR. A lower level of security means lowering the FAR and FRR. A higher level of security means increasing the FAR and FRR.

Trilogy Access-Control Locking System Features

Trilogy DL2800 Economical Audit Trail Pushbutton Lock:

- Rugged clutch mechanism ensures long life and durability.
- 150 scheduled events, including four "quick schedules" for programming the four most common time schedules in one step.
- 200 user codes including master, manager, supervisor, and basic users; also, one-time entry code.
- Weather-resistant performance with an operational temperature range of +151 to −20°F.
- Times entry allowance from 5 to 15 seconds.
- Greater security with temporary keypad lockout after three unsuccessful code-entry attempts.
- Audit trail with time/date stamp for a printable record of all electronic activity.
- Real-time clock allows logging of events to within 1 second of accuracy.
- Relay may be programmed to energize when one or more selected events occur.
- Four time-out functions allow a user to enable other users or unlock the lock for some time period without needing to return to the lock it.

Figure 10.8 The Trilogy DL200 Pushbutton lock. (*Courtesy of Napco Security Systems.*)

- 1000-event log
- Easy to install; retrofits most cylindrical locksets and digital locks.
- Battery-operated; uses five AA batteries.
- Available in standard key override and interchangeable core models accepting six- or seven-pin cores from Arrow, Best, Falcon, or KSP; other 1C core preps available.
- Available in standard Trilogy lever set (straight) or with new Regal (curved) lever. See figure 10.8.

11

Safety and Security Lighting

Lighting is not only a low-cost form of security, but it also can help to prevent accidents, create moods, and enhance the beauty of any home. This chapter shows how you can make the best use of lighting inside and outside your home. A dark house is an invitation to crime and creates a high risk for accidents. When you approach your home late at night, you need to be able to walk to your entrance without tripping over something—or *someone*—in your path. When you're inside, you need to be able to move from room to room safely. Your home should be well lighted on the inside, in the areas directly outside the doors, and throughout the yard (Figure 11.1). *Well lighted* doesn't necessarily mean a lot of light; it means having the light sources and controls in the right places.

Light Sources

Our most common light source is the sun, which we cannot control. We have artificial light sources available for use at night and in some indoor locations during the day. Important differences among artificial sources include color, softness, brightness, energy efficiency, and initial cost.

Light sources you might consider for home use include standard incandescent, halogen, fluorescent, and high-intensity discharge (HID) lighting. The HID family of lighting includes low-pressure sodium, high-pressure sodium, mercury-vapor, and metal halide lights.

An *incandescent* light source relies on heat to produce light. The standard bulbs used in most homes are incandescent (lighting designers call them *A-lamps*). They have a metal filament that is heated by electricity. Standard incandescent bulbs are inexpensive, readily available, and suitable for most home fixtures. They light up almost immediately at the flip of a light switch. However, using heat to produce light isn't energy efficient; in the long run, incandescent lighting can be more costly than other sources that require special fixtures.

Figure 11.1 Good outdoor lighting is well-balanced lighting that lets you see who's at the door at night. (*Courtesy of Baldwin Hardware Corporation.*)

Halogen, a special type of incandescent source, is slightly more energy efficient than standard incandescent lighting. A halogen bulb uses a tungsten filament and is filled with a halogen gas.

Fluorescent lighting uses electric current to make a specially shaped (usually tubular) bulb glow. The bulbs come in various lengths, from 5 inches to about 96 inches, and they require special fixtures. You might not want to use fluorescent lighting with certain types of electronic security systems because it can interfere with radio reception. Nor would you ordinarily use fluorescent lighting outdoors in cold climates: it's very temperature sensitive.

For outdoor lighting, you might use high-intensity discharge (HID) sources, which are energy efficient and cost little to run for long periods of time. Like fluorescent lighting, HID sources require special fixtures and can be expensive initially. Another potential problem with HID sources is that they can take a long time to produce light after you've turned them on. Startup time can be unimportant if you use a light controller to activate the lights automatically when necessary.

Light Controllers

Timers are among the most popular types of controllers for indoor and outdoor lighting. The newest timers can do much more than just turn lights on and off at preset times. Programmable 24-hour wall switch timers, for example, will turn your lights off and on randomly throughout the night and early morning. This feature is useful because many burglars will watch a home to see whether the lights come on at the exact same time each night—an indication that the home is empty and a timer has been preset. Another feature of some new timers is built-in protection against memory loss. After a power failure, they "remember" how you had programmed them. Some models adjust themselves automatically to take into account daylight savings time changes. Most timers used by home-owners cost between $10 and $40.

Another low-cost way to control lights is with sound or motion sensors. You can buy one of these sensors and connect it to, say, a table lamp in your living room. When you (or someone else) walk into your living room at night, the light will come on automatically.

Some floodlights come with a built-in motion sensor. If you install them out-side at strategic places, they will warn you of nighttime visitors. You might install one facing toward your driveway, for instance, so that it will light up the area when a car pulls in. Floodlights generally sell for less than $50.

Preventing Accidents

To prevent accidents at nighttime, you need to be able to see potential hazards. When walking down a flight of stairs, for instance, it's important to be able to see whether any objects are in your way. In too many homes there is a need to stumble through dark areas or to grope for a light switch or a series of light switches before reaching the bathroom or kitchen.

Can you use motion-activated sensors to avoid this problem? You'd probably need a lot of them to cover every path you might take at night. Simpler and more convenient options are available.

One useful practice is to install night lights near your light switches so that you can reach them more easily. Night lights cost only a few dollars each, and they consume very little power. Another option is to use three-way switches. They allow you to turn a light on and off at more than one location, such as at the top and bottom of a flight of stairs.

The most convenient way to use lighting indoors is to tie all (or most) of the switches into an easily accessible master control panel. Then you could turn on a specific group of lights by just pushing a button. One button could turn on a pathway of lights from, say, your bedroom to the bathroom. By pushing another button, you could activate a pathway of lights from your bedroom to the kitchen. Some or all of your outdoor lights also could be tied into your master control system.

Installing Motion-Activated Outdoor Lighting

You can install a motion-activated light virtually anywhere indoors or outdoors—on the side of your home, on your porch, in your garage, or wherever it's needed. Many of these lights are simple two-wire installations in which the hardest part of the process is the proper positioning of the lighting unit. Models are available in various colors and styles to match your decor.

120-Volt Lighting

A 120-V lighting system will provide brighter light than a low-voltage system. The brighter light may be especially useful outside if you need to illuminate a large area. Installing a 120-V system is more involved and the materials are more expensive than those used in a low-voltage system.

Before beginning the installation, familiarize yourself with your local electrical code, and obtain any required permits. You may have to draw up a plan and have it reviewed by your local building inspector before you're allowed to install 120-V lighting.

You'll need to decide what materials to use—receptacles, cables, switches, boxes, conduit, conduit fittings, wire connectors, and so on. Your local code already may have made some of these decisions for you. You may be required to use rigid metal conduit rather than PVC conduit, for instance, or you may be restricted to using only certain types of wire.

Home and Office Lighting Checklist

1. Keep the areas outside your exterior doors and windows well lit at night.
2. Make sure your outside lighting leaves no shadows that will provide cover for burglars (multiple light sources prevent shadows better than a single light source).
3. Use weather-resistant protective covers on outside lights.
4. Immediately report any nonworking streetlight.
5. Keep blinds and drapes closed at night.
6. Either leave some lights on all night or use automatic times near stairs, hallways, and bathrooms.
7. Use timers to turn lights on and off in various rooms whenever you're away at night.

Expert Advice

I asked lighting designer Julia Rezek for her secrets to choosing and installing residential lighting. She began working as a lighting designer in 1983, after graduating from UCLA. Ms. Rezek has taught lighting classes at UCLA's University Extension Program and at the Otis Parson Institute of Design in Los Angeles. She is the Director of Residential Lighting for Grenald Associates in Culver City, California, and is a member of the International Association of Lighting Designers.

Q. What suggestions would you give a homeowner who wants to use lighting for security but doesn't want the lighting to be harsh and offensive?

A. I recommend installing two complementary types of lighting systems. One is for panic situations—such as when you're awakened by a noise at night—to turn on very bright and offensive floodlights in your yard or driveway. The other lighting system should be much softer and be connected to a time clock so it can come on automatically every night.

Q. Which lighting sources and fixtures do you recommend for indoor lighting?

A. When planning the lighting for an indoor space, I consider what activity the space will be used for, what effect is desired from the light source, and how the lighting should be controlled. A primary concern for indoor lighting is to avoid glare and create an environment that is soft, comfortable, and easy to control.

   Most homeowners tend to use only standard incandescent light bulbs, but such light is yellow and deficient in the blue-green.

A. First, anyone installing outdoor lighting should know how to comply with all electrical codes, which vary from one city to another. A typical installation would involve installing transformer boxes somewhere in the back of the house, around the corner, or somewhere that they can't be seen. Then the homeowner would need to dig a trench about 6 inches deep to run flexible wire to the different light sources. Although it isn't usually necessary to run the wire through rigid conduit, it's a good idea to run it through PVC piping to prevent a lawn mower from damaging the wire.

Q. What lighting sources do you recommend for porches and stoops?

A. Those are transition areas where people are coming from the outside to the inside. You want those areas to have a warm and yellow quality, so your guests will feel drawn toward the warmth of light and will know that inside your home is also warm and inviting.

Closed-Circuit Television Systems

Locks and other physical security devices can be more effective when used in conjunction with a closed-circuit television (CCTV) system. Such a system can allow several areas, such as elevators, entrances and exits, parking lots, lobbies, and cash-handling areas, to be monitored constantly. Such monitoring can deter crime and reduce a person's or company's security costs.

Imagine having an extra pair of eyes that can be several places at once. Without getting out of bed, you would be able to see who's at your front door, who's walking around your yard, and who's driving onto your property. You also could keep an eye on any room of your home or business. With a CCTV system, you can do all these things and more.

This chapter looks at the various components that make up a system and shows various ways that such a system can be used.

Tips for Using a CCTV System in Your Home or Office

1. For a home or office, a CCTV system is used primarily to monitor who comes through the front or back door. If children are there, you also might want to install cameras in playrooms.
2. It's important for the camera to be positioned so that the light is between the camera and the scene being viewed. Make sure that the light doesn't come in and blind the camera. Too much direct light will result in a poor picture on the monitor.
3. Install monitors in several readily accessible places, such as the master bedroom, the kitchen, and the family room.
4. A pan and tilt can be useful, but generally, it isn't necessary if all you want to do is see who's coming to your door. By using the proper lens, you usually can see a wide enough area.
5. Most home and office CCTV systems are overt. Covert cameras are useful if you think overt cameras are ugly or if you want to discreetly monitor babysitters.
6. Don't install covert cameras in places where people have a reasonable expectation of privacy, such as in a bathroom.
7. If you need a state-of-the-art system or are building a new home, you should use a professional CCTV system installer.

How CCTV Systems Work

A CCTV system simply transmits images to monitors that are connected to the system's camera. The system's basic components are a video camera and monitors connected to it by coaxial cable. This type of installation wouldn't be very useful for security purposes: You would have to prop the camera up in a room, point it to a fixed location that you wanted to protect, and then go and stare at the monitor.

For security, you need a camera that works while you're not around, can be controlled from a remote location, and can be connected to a burglar alarm system. If you need to monitor more than one location—for example, your front door *and* your back door—you may want to be able to use one monitor that is receiving images from both cameras. You also may want the option of monitoring both cameras at once. All these and many other features are possible with CCTV systems that are currently selling for less than $400 (not including installation).

When you know what's available, you can choose the cameras, monitors, and optional components that will create a custom CCTV system within your budgeted amount. If the system isn't too complex, you'll probably be able to install it yourself. Most of the CCTV systems used by homeowners are easier to install than a hardwired alarm system.

Cameras

Two types of cameras are used commonly in CCTV systems: the tube camera, the older type, which is fast becoming obsolete, and the closed-coupled-device (CCD) camera, which lasts longer and works better. CCD cameras cost a little more than tube cameras, but they have been coming down in price steadily, whereas the price of tube cameras has remained the same. With the demand for CCD cameras continuing to outpace the demand for tube cameras, many camera makers are discontinuing their line of tube cameras.

Cameras come in color and black-and-white transmission models. A color camera requires maximum and constant light to be able to view a scene properly and shouldn't be used outdoors or in any area that sometimes gets dark. A black-and-white camera is more tolerant of all types of lighting conditions and is less expensive.

The choice between color and black-and-white transmission is usually simple. In virtually every residential situation, black-and-white transmission is more cost effective and much less troublesome. Color cameras are needed only in banks and at other surveillance sites where the cameras' videotapes may become evidence in a court case.

Cameras come in many sizes, described by their lens diameter. The three most common sizes are $1/3$, $1/2$, and $2/3$ inch. The $2/3$-inch camera covers *more* area and gives better resolution than the $1/3$-inch camera. Generally, the larger the camera, the better is the picture.

Monitors

In many ways, your choice of a monitor is as important as your choice of a camera. The quality of the picture you receive on your monitor depends on both. The camera and monitor work together, much like speakers and an amplifier in an audio system. If you have a great amplifier with poor speakers or great speakers with a poor amplifier, you'll get poor sound because the sound is filtered through both devices before you hear it. In a CCTV system, the picture is filtered through both the camera and the monitor before you see it.

For transmission of a color image, your monitor and your camera must be color equipment. If either is black-and-white, you'll receive a black-and-white picture. Monitors, described based on their screen diameter, range in size from about 4 inches to over 21 inches. The most common monitors for home use are the 9- and 12-inch sizes.

To save money, you can buy a radio frequency (RF) modulator for your television and convert it to a monitor. Then you'll be able to view the camera's visual field just by turning your television to a particular channel (usually either channel 3 or channel 4). You can buy an RF modulator for under $50.

Peripheral Devices

One of the most popular devices for CCTV systems is a pan-and-tilt unit. It gives a camera the ability to tilt up and down and to rotate up to 360 degrees left to right or right to left. With a pan-and-tilt unit, you will be able to zero in on items (and people) within a wider camera range. By using a pan-and-tilt unit in a large installation, you'll need only one camera, not several.

Pan-and-tilt units have long been used in airports, banks, and other large commercial installations. Because the units often cost over $1000, they're rarely included in home CCTV systems.

Another complementary device you can use with your CCTV system is a sequential switcher. With one monitor, the switcher will allow you to receive pictures from several cameras. You can watch one camera field for a while and then switch over to another.

If you want a continuous record of what the camera sees, you can install a video lapse recorder, which will span up to 999 hours with individual photo frames on one standard VHS 120 videotape. If you wish, the current time and date can be recorded automatically on each frame.

If you want to record only unwanted persons who enter a particular area or room, you can use a camera that has a built-in motion-detecting capability and is connected to an alarm system. The alarm will be triggered when the camera begins taping.

Another option available with today's CCTV systems is the dual-quad unit, which gives a standard monitor the capability of showing as many as 16 pictures at one time from 16 separate cameras. Dual-quad units can cost anywhere from $1000 to $15,000.

Installing a CCTV System

The specific installation methods that are best for you will depend on the components you've chosen and how you want to use them. Many of the hardwiring methods for installing burglar alarms, detailed in Chap. 9, are useful for installing CCTV systems.

Most CCTV systems can be installed either independently or incorporated into a burglar alarm system. If you tie the CCTV system into a burglar alarm system and use a videotape recorder, you can set the CCTV system to begin recording automatically whenever the alarm is triggered. You also can have your system record sounds.

Your installation can be either overt or covert. Most homeowners use an overt system because they want would-be burglars to know that they are being watched. A camera that's prominently connected to the side of your home certainly would act as a deterrent.

Some people consider cameras inside a home unattractive and threatening, but there are some advantages to keeping the cameras out of sight. Hidden cameras allow you to make a secret videotape of a burglar.

To install a covert system, you'll need to buy small cameras specially designed to be installed in the corner of a wall or in a wall cutout. They sell for between $100 and $200. Some covert cameras are disguised to look like clocks and other common objects. Their prices start at about $200.

You should seek legal advice before installing a covert system; in some jurisdictions, such systems are illegal. Some states consider secret audiotaping to be in violation of wiretap laws.

Video Intercoms

A video intercom system is a CCTV system that lets you talk to the people you're seeing through its camera. With many models, you can choose to see and hear a person without the person knowing you're home. Like burglar alarms, video intercoms can give your home a high level of security—and they're usually easier to install than burglar alarms.

In some video intercom systems, the cameras, monitors, intercoms, and peripherals are all separate components. Other systems have integrated components, such as a monitor with a built-in intercom or a camera with built-in peripherals. Because they have fewer components, integrated units usually are simpler to install. They also tend to take up less space and look nicer. The main problem with most integrated systems is that they can't be expanded to add on sophisticated peripherals.

Some models are designed to incorporate a variety of peripherals. Aiphone's Video Sentry Pan Tilt, for example, includes an integrated camera, a monitor, an intercom, and a motorized pan-and-tilt unit (Figure 12.1). Its camera can scan 122 degrees horizontally and 76 degrees vertically—up to four times the area visible with a fixed camera. A button on the monitor unit allows you to control the panning and tilting actions of the camera unit.

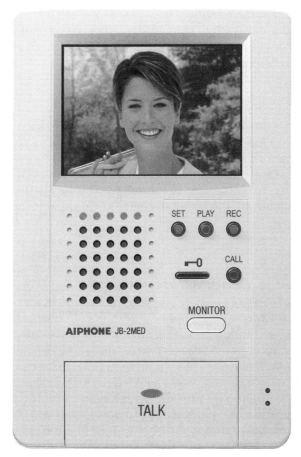

Figure 12.1   Aiphone's hands-free color video, master station with picture memory.

Separate components allow more flexibility during installation. You can mount the outside intercom where visitors can reach it easily, for instance, and place a separate camera where they can't see it. Separate cameras can be installed easily on gates, near swimming pools, and at other outside locations.

Monitors range in size (based on the diameter of the screen) from 4 to about 20 inches. Most integrated units have a built-in 4-inch black-and-white monitor. The 9- and 12-inch sizes are popular for monitors used with separate units.

Lighting Considerations

Different cameras need different amounts of light to view a scene properly. Whether it comes from the sun, from starlight, or from light bulbs, the amount of light needed is measured in lux. The fewer lux a camera needs, the more adaptable it is to nighttime viewing. Most color cameras need a minimum illumination of 3 lux; black-and-white cameras usually need only 1 lux. Some security units require 0 lux because their cameras have built-in infrared diodes that produce the necessary amount of light.

Buying Tips

You'll find the lowest-priced video intercoms at large hardware and builders' supplies stores. Many low-cost models are clones of expensive popular brands.

To get the best price on a brand-name model, ask the manufacturer for product literature and for the phone numbers of local distributors. Call several distributors to get their current prices. (Be sure that the quotes include shipping charges.) Many distributors actively compete with one another, frequently changing their prices.

Installation Tips

Is there a "best way" to install a video intercom? The installation method will depend on the model you've chosen, where you want to use it, and how you want to use it. Most manufacturers will supply you with installation instructions, templates, and a wiring diagram. The following tips may make the job easier.

In a typical installation, you'll first need to decide where to mount the camera and the monitor units. Regardless of its minimum lux requirement, the camera should be placed where there is always enough light for you to read a book page (a porch light or streetlight may provide enough light at nighttime). With less light, you probably won't be able to see people well on your monitor screen. Don't position the camera so that it will be subjected to direct, glaring sunlight. Another consideration is that the place you choose should allow for easy wiring access to the indoor monitor.

The monitor unit can be placed in any location that's convenient for you (often in the room where you tend to spend the highest percentage of your waking time) and near an electrical outlet. It's usually best to run your cable (you'll be using coaxial two- or four-wire cable) before mounting the camera and monitor units. In that way, if you have trouble getting the cable to the desired locations, you can choose other mounting locations without leaving unsightly screw holes.

13

Safes and Vaults

Virtually every consumer and business has documents, keepsakes, collections, or other valuables that need protection from fire or theft. But most people don't know how to choose a device that meets their protection needs, and they won't get much help from salespeople at department stores or home-improvement centers. By knowing the strengths and weaknesses of various types of safes, you will have a competitive edge over such stores.

No one is in a better position than a knowledgeable locksmith to make money selling safes. Little initial stock is needed, they require little floor space, and safes allow for healthy price markups. This chapter provides the information you need to begin selling safes to businesses and homeowners.

Types of Safes

There are two basic types of safe: fire (or record) and burglary (or money). *Fire safes* are designed primarily to safeguard their contents from fire (Figure 13.1), and *burglary safes* (Figure 13.2) are designed primarily to safeguard their contents from burglary. Few low-cost models offer strong protection against both potential hazards. This is so because the type of construction that makes a safe fire resistant—thin metal walls with insulating material sandwiched in between—makes a safe vulnerable to forcible attacks. The construction that offers strong resistance to attacks—thick steel walls—causes the safe's interior to heat up quickly during a fire.

Most fire/burglary safes are basically two safes combined, usually a burglary safe inside a fire safe (Figure 13.3). Such safes can be very expensive. If a customer needs a lot of fire and burglary protection, you might suggest that he or she just buy two safes. To decide which type of safe to recommend, you need to know what your customer plans to store in it.

Figure 13.1 Fire safes. (*Courtesy of Gardall Safe Corporation.*)

Figure 13.2 Burglary safes. (*Courtesy of Gardall Safe Corporation.*)

Figure 13.3 A combination fire/burglary safe is two safes in one. (*Courtesy of Gardall Safe Corporation.*)

Safe Styles

Fire and burglary safes come in three basic styles based on where the safe is designed to be installed. The styles are wall, floor, and in-floor. *Wall safes* are easy to install in homes and provide convenient storage space (Figure 13.4). Such safes generally provide little burglary protection when installed in a drywall cutout in a home. Regardless of how strong the safe is, a burglar simply can yank it from the wall and carry it out. To provide good security, not only does a wall safe need a thick steel door, but it also needs to be installed with concrete in a concrete or block wall.

Floor safes are designed to sit on top of a floor. Burglary models should either be over 750 pounds or be bolted in place. Figure 13.5 shows some popular floor safes. One way to secure a floor safe is to place it in a corner and bolt it to two walls and to the floor. (If you sell a large wall safe, make sure that your customer knows that the wheels should be removed from the safe.)

In-floor safes are installed below the surface of a floor (Figure 13.6). Although they don't meet construction guidelines to earn an Underwriters Laboratories (UL) fire rating, properly installed in-floor safes offer a lot of protection against fire and burglary. Because fire rises, a safe below a basement floor won't get hot inside quickly. For maximum burglary protection, the safe should be installed in a concrete basement floor, preferably near a corner. Such placement makes it uncomfortable for a burglar to attack the safe.

Make sure that your customer knows that he or she should tell as few people as possible about the safe. The fewer people who know about a safe, the more security the safe provides.

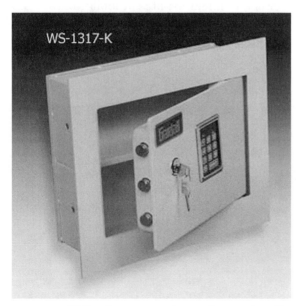

Figure 13.4 Wall safes are convenient and easy to install.

Figure 13.5   Floor safes.

Installing an In-Floor Safe

Although procedures differ among manufacturers, most in-floor safes can be installed in an existing concrete floor in the following way:

1. Remove the door from the safe, and tape the dust cover over the safe opening.

2. At the location where you plan to install the safe, draw the shape of the body of the safe, allowing 4 inches of extra width on each side. For a square-body safe, for example, the drawing should be square, regardless of the shape of the safe's door.

3. Use a jack hammer of a hammer drill to cut along your marking.

4. Remove the broken concrete, and use a shovel to make the hole about 4 inches deeper than the height of the safe.

Figure 13.6   In-floor safes.

5. Line the hole with plastic sheeting or a weatherproof sealant to resist moisture buildup in the safe.

6. Pour a 2-inch layer of concrete in the hole, and level the concrete to give the safe a stable base to sit on.

7. Place the safe in the center of the hole and shim it to the desired height.

8. Fill the hole with concrete all around the safe, and use a trowel to level the concrete with the floor. Allow 48 hours for it to dry.

9. After the concrete has dried, trim away the plastic and remove any excess concrete.

In-floor safes are usually installed near walls to make them hard for burglars to use tools on them.

To install an in-floor safe in concrete, use a jackhammer to cut a hole in the concrete about 4 inches wider than each side of the safe, and make the hole 4 inches deeper than the safe's height. Line the hole with plastic, and thinly line the bottom of the hole with a level layer of concrete to give the safe a solid base.

Moving Safes

Getting a heavy safe to your customer can be backbreaking unless you plan ahead. Consider having the safe drop shipped, if that's an option. Most suppliers will do that for you.

As a rule of thumb, have one person help for each 500 pounds being moved. If a safe weighs more than a ton, however, use a pallet jack or machinery mover. When moving a safe, never put your fingers under it. If the safe has a flat bottom, put three or more 3-foot lengths of solid steel rods under it to help side the safe around, and use a prybar for leverage.

Special Safe Features

Important features of some fire and burglary safes include relocking devices, hardplate, and locks. Relocking devices and hardplate are useful for a fire safe but are critical for a burglary safe. If a burglar attacks the safe and breaks one lock, a *relocking device* automatically moves into place to hold the safe door closed. *Hardplate* is a reinforcing material strategically located to hinder attempts to drill the safe open. Never recommend a burglary safe that doesn't have relockers and hardplate.

Safe locks come in three styles: key-operated, combination dial, and electronic. *Combination dial* models are the most common. They are rotated clockwise and counterclockwise to specific positions. *Electronic* locks are easy to operate and provide quick access to the safe's contents. Such locks run on batteries that must be recharged occasionally. For most residential and small-business purposes, the choice of a safe lock is basically a matter of personal preference.

Underwriters Laboratories Fire Safe Ratings

UL fire safe ratings include 350-1, 350-2, and 350-4. A 350-1 rating means that the temperature inside the safe shouldn't exceed 350°F during the first hour of a typical home fire. A safe rated 350-2 should provide such protection for up to 2 hours of fire. Safes with a 350-class rating are good for storing paper documents because paper chars at 405°F. Retail prices for fire safes range from about $100 to over $4000. Most models sold in department stores and home-improvement centers sell for under $300.

Most fire safes aren't designed to protect computer disks, DVDs, CDs, or videotape. To store computer disks and other magnetic media, you need to use a media record safe. When it's up to 1832°F outside, a media safe keeps the interior below 125°F. Figure 13.7 shows a media safe.

Figure 13.7 A media safe. (*Courtesy of Gardall Safe Corporation.*)

Underwriters Laboratories Burglary Safe Standard

The UL Standard 689 is for burglary-resistant safes. Classifications under the standard, from lowest to highest, include Deposit Safe, TL-15, TRTL-15x6, TL-30, TRTL-30, TRTL-30x6, TRTL-60, and TXTL-60. Figure 13.8 shows depository safes. The classifications are easy to remember when you understand what the sets of letters and numbers mean. The two-set letters in a classification (TL, TR, and TX) signify the type of attack tests a safe model must pass. The first two numbers after a hyphen represent the minimum amount of time the model must be able to withstand the attack. An additional letter and number (e.g., x6) tells how many sides of the safe have to be tested.

The TL in a classification means that the safe must offer protection against entry by common mechanical and electrical tools, such as chisels, punches, wrenches, screwdrivers, pliers, hammers and sledges (up to 8-pound size), and pry bars and ripping tools (not to exceed 5 feet in length). TR means that the safe also must protect against cutting torches. TX means that a safe is designed to protect against cutting torches and high explosives.

For a safe model to earn a TL-15 classification, for example, the sample safe must withstand an attack by a safe expert using common mechanical and electrical tools for at least 15 minutes. A TRTL-60 safe must stand up to an attack by an expert using common mechanical and electrical tools and cutting torches for at least 60 minutes. A TXTL-60 safe must stand up to an attack with common mechanical and electrical tools, cutting torches, and high explosives for at least 60 minutes.

In addition to passing an attack test, a safe must meet specific construction criteria before earning a UL burglary safe classification. To be classified as a depository safe, for example, the safe must have a slot or otherwise provide a means for depositing envelopes and bags containing currency, checks, coins, and the like into the body of the safe, and it must provide protection against common mechanical and electrical tools.

Figure 13.8 Depository safes.

The TL-15, TRTL-15, and TRTL-30 safe either must weigh at least 750 pounds or must be equipped with anchors and instructions for anchoring the safe in a larger safe, in concrete blocks, or to the premises in which the safe is located. The metal in the body must be the equivalent to solid open-hearth steel at least 1 inch thick having an ultimate tensile strength of 50,000 pounds per square inch (psi). The TRTL-15x6, TRTL-30x6, and TRTL-60 must weigh at least 750 pounds, and the clearance between the door and jamb must not exceed 0.006 inch. A TXTL-60 safe must weigh at least 1000 pounds.

Safes with the TL-15 and TL-30 ratings are the most popular for business uses. Depending on the value of the contents, however, a higher rating may be more appropriate. Price is the reason few companies buy higher-rated safes. The retail price of a TL-30 can exceed $3500. A TXTL-60 can retail for over $18,000.

Such prices cause most home owners and many small businesses to choose safes that don't have a UL burglary rating. When recommending a nonrated safe, consider the safe's construction, materials, and thickness of door and walls. Better safes are made of steel and composite structures (such as concrete mixed with stones and steel). Safe walls should be at least $1/2$-inch thick and the door at least 1 inch thick. Make sure that the safe's boltwork and locking mechanisms provide strong resistance to drills. Another type of fire safe is a gun safe. It's made large enough to hold guns (Figure 13.9).

Figure 13.9 A gun safe. (*Courtesy of Gardall Safe Corporation.*)

Buying Safes

Safes vary greatly in quality, utility, and price. Even with your special knowledge, you need to buy high-quality safes at the lowest prices possible because you're also competing against other locksmiths. You need to decide whether to buy directly from a manufacturer or a distributor. The advantage of buying from a manufacturer is a lower price when ordering volume. Most locksmith shops don't purchase enough safes to benefit from the potential savings. The advantages of dealing with a distributor are the option of buying in small quantities and (usually) faster delivery.

Although it isn't necessary to stock a lot of safes, you'll need to have a few models on display if you want to be taken seriously as a safe seller. Most people want to see a product before buying it. Initially, you shouldn't stock large, heavy safes because they're expensive, hard to transport, and usually don't sell quickly.

Automobile Security

In 2004, there were an estimated 1,237,114 motor vehicle thefts in the United States. According to the Insurance Information Institute, a motor vehicle is stolen every 26 seconds in the United States.

Some vehicles are stolen by juveniles for joy riding, but the greatest threat to your car is professional car thieves who search for cars to sell or strip for parts. Using special tools and methods, within minutes they can take whatever model they need.

Expensive and late-model cars aren't the only ones the professional thieves want. The average value of a vehicle at the time of theft is only about $6000. When sold through a "chop shop," some car parts (e.g., engines, doors, hoods, fenders, and so on) can be more valuable than an intact car.

There are many precautions you can take, and several low-cost devices are available to keep thieves out of your car. Some highly advertised antitheft devices only make car thieves laugh. This chapter is about which devices work, which don't, and why.

Antitheft Devices

The four basic types of antitheft devices for vehicles are cutoff switches, supplemental locks, alarms, and stolen-vehicle retrieval systems. Cutoff switches disable a car that someone has tried to start without using the right key. Some cutoff switches allow a car to run for a short while; others prevent the engine from starting. Those that allow the car to run can be dangerous not only to the car thief but also to others on the road.

Supplemental locks are used for securing various parts of a car—doors, trunks, and steering wheels. The most popular type is the steering wheel lock, which is sold under many brand names. One style is a cane-shaped bar that locks the steering wheel to the brake pedal. Another style uses the dashboard as an obstruction to prevent the steering wheel from turning. Most steering wheel locks are easy to saw off, pick open, or bypass.

When a steering wheel lock is used properly, however, it lets thieves know that the car is owned by a security-conscious person and may make them wonder what other security measures have been taken. Any car thief would prefer to "hit" a car that doesn't have a steering wheel lock in place. Unfortunately, thieves are more likely to find a steering wheel lock in a car's trunk than on its steering wheel because many owners find the constant need to put the lock on and take it off too much of a nuisance.

Car Alarms

Some alarms can be useful for deterring and catching car thieves, but many models are virtually worthless. There are two basic types of car alarms. A *passive alarm* arms itself automatically, usually when the car is turned off and the last door is closed. An *active alarm* requires you to do something special—push buttons or flip a switch—to arm the car. Because it's easy to forget to arm a system, the passive systems provide more security (Figure 14.1).

Many insurance companies give owners of cars with a passive system a 5 to 20 percent discount on the comprehensive portion of their car insurance. The states that require insurance companies to do so are Illinois, Kentucky, Massachusetts, Michigan, New York, and Rhode Island.

Here's how a typical car alarm works. The alarm consists of detection devices, a siren or other warning device, and a control module. All the components are connected to one another by wire or radio waves to form a complete electric circuit. The detection devices are strategically placed so that a thief can't take the car without causing a break in the circuit. When a door is opened, for instance,

Figure 14.1   Car alarm components. (*Courtesy of Black Widow Security, Inc.*)

a circuit break should occur, causing the sirens, lights, and other warning devices to activate.

Most car alarms have current-detection devices; they sense the current drop that occurs when a car's courtesy light comes on. This feature alone usually isn't enough to protect a vehicle because a thief can gain entry in ways that don't activate the courtesy light. Other types of detection devices respond to the sound of breaking glass, to vibrations, or to motion within the car. Generally, the more types of detection devices a system uses, the more effective it will be.

The most common detection device is the pin switch, which senses the opening of a door or trunk. The thin, spring-loaded switch is installed in a small hole in the frame around the door (or trunk) so that it is compressed when the door is closed. While it's compressed, the switch helps to complete an electric circuit. When the door is opened, the switch springs up, breaking the circuit. The switch has a single wire; the car's metal chassis is the ground portion of the circuit.

A motion detector can be used to protect against thieves who use glass cutters to gain entry because it can sense cut glass falling into the car. It can be especially useful for vehicles that have vent windows or sliding windows. Some alarm systems use strobe lights or a car's headlights or parking lights as warning devices. A light-emitting diode (LED) installed on a door or dashboard also can be used as a warning device. It flashes continuously when the alarm is armed, letting thieves know that the car is protected. It flashes at a faster rate whenever the alarm has been activated, letting owners know the car has been tampered with.

The most popular type of warning device is a siren. Many models sound at 100 or 113 decibels; better models sound at a minimum of 120 decibels. Some locales have restrictions on the types of sirens that may be used. You should use one that automatically shuts off and resets itself after a certain period of time. A siren that blasts continuously will drain your battery and be a nuisance to people nearby.

The control module of a car alarm is the brain of the system. When it learns from a detection device that the electric circuit has been broken, the control module decides when to activate the sirens and other warning devices. It usually has an entry/exit delay feature to allow you enough time to get into and out of your car without triggering the alarm. Depending on the model, you can adjust the delay time within 12 to 40 seconds. The shorter the delay, the less time a thief will have to disable the system before the alarm is triggered. No delay should be set for the trunk and hood.

The biggest problem with most car alarms is that they depend on passersby to intervene. Car thieves know that if they can disarm an alarm quickly, they will attract little or no attention from passersby. Even if someone does challenge the car thief, all he or she has to say is that it's his or her car. Few people care enough about strangers' cars to risk offending someone who may or may not be a thief.

Rather than relying on passersby to tell a thief to leave your car alone or to call the police, you can carry a remote pager in your pocket. It will warn you whenever your car alarm is activated. Some models allow remote arming and disarming of the alarm system.

The advertised range of a pager is often the maximum range that can be achieved under the best conditions. The actual range you need will depend on where your car is and how far away from it you'll be. If your car is in an underground parking garage and you're in a high-rise building, for instance, the pager will have less range than if you and the car were in the same open field. The more obstructions between you and the vehicle, the less range the pager will have. Good pagers have a range of over 2 miles in most situations.

Special Features in Car Alarms

Car alarms are no longer designed only for security. Many offer a variety of conveniences. Some car alarms allow you to use a remote control to start or shut down your car's engine, heater, and air conditioner; lock and unlock the doors; open the trunk; control windows and a sunroof; activate the interior courtesy light; flash the parking lights; arm and disarm the alarm; monitor and recharge the car's battery; and control electric garage doors.

Three special security features to look for when choosing an alarm are a self-recharging backup battery, a hood lock, and antiscanning circuitry. A backup battery will replace your primary battery if it runs down or is disconnected by a thief. A hood lock connected to your alarm will activate the system if a thief tries to get under the hood. (The factory-installed hood locks on most cars are easy to defeat.)

Before buying an alarm that allows remote-control operation, make sure that it has antiscanning circuitry. Hand-held frequency scanners can be made from parts found in any electronic hobby shop or bought by mail order. Car thieves use these scanners for rapid transmission of different codes until one is found that will disarm a car's alarm. Alarm systems with antiscanning circuitry can detect the use of a scanner and resist being controlled by it.

Warranties

Most car alarms come with a warranty. Don't take the warranty for granted: You may need service or a replacement. All car alarm manufacturers occasionally make defective products.

Carefully read the warranty to be sure of what it does and doesn't cover. A typical warranty from a manufacturer covers parts and labor for one year. It usually applies only to defective products sold to the original retail purchaser and doesn't cover problems caused by a faulty installation.

Because rights covered by warranties vary from state to state, you may have more rights than the manufacturer has listed. For example, although most warranties say that they exclude or limit "incidental or consequential damages," some states don't allow such an exclusion or limitation.

The dealer who sells you an alarm or the company that installs it also may offer a warranty. If you're not certain about your rights under a warranty, consult with your attorney.

Factory versus Custom Installation

Over twenty-six car lines have alarms as standard equipment. Many others provide them as options. The big problem with such alarms is that they're standardized. A thief needs only one owner's manual to learn how to disable all factory-installed alarms in a certain make or model of car.

A custom-installed system has the element of surprise. The thief won't know what type is installed, what special security features are included, or where the various components have been placed. Many custom systems cost less and are more sophisticated than factory-installed alarms.

As with home alarms, don't use a window sticker that identifies the brand or model of your system. Alarm makers like the free advertising, but the more that is known about your system, the easier it will be to defeat. Buy a generic window sticker or use one from a different alarm system.

Choosing an Installer

Will you need a professional installer? This depends on the system you choose and your ability to wire. The more components and wires a system has, the harder it will be to install. The easiest alarms to install are portable, self-contained models that have a built-in current sensor, a motion detector, and a siren.

Most alarms are placed on a car's dashboard and installed by plugging a cord into the cigarette lighter or connecting a single wire to a fuse box. A thief can disable such a unit within seconds just by unplugging the cord and removing the backup battery. Better portable models are installed under a car's hood and connected to the car's battery by a single wire.

Some manufacturers of high-tech systems provide installation instructions only to factory-authorized dealers and won't warrant a product installed by anyone else. With most systems, however, you can choose your own installer.

Car alarms are installed by burglar alarm companies, locksmiths, auto repairs shops, and audio equipment service centers. Anyone who can wire a stereo system in a car can wire most car alarms, but proper installation of a car alarm requires security considerations that don't apply to stereo equipment. The alarm components should be placed and wired in ways that will confuse sophisticated car thieves. If you want your system installed professionally, look for a company that specializes in car alarms or other security systems.

Find out how long the company has been installing car alarms. Ask for names and phone numbers of previous customers, and call those references to find out what problems they may have had with their systems. If any of them is unhappy with the installer, look for another installer. Their unfavorable opinion indicates that the installer didn't obtain their permission to be contacted by you—a telltale sign of an installer who has little respect for security.

Call your local Better Business Bureau or your state's attorney general's office and ask whether complaints have been filed against the company. Find out how long the installer has been doing business in your area.

After finding two or three acceptable installers, compare their prices and warranties. It's not unusual to find an installer that charges up to twice as much as others in the same area.

Stolen-Vehicle Retrieval Systems

The latest advances in automobile security are stolen-vehicle retrieval systems. The recovery rate for stolen vehicles equipped with retrieval systems is about 95 percent, compared with only 64 percent for vehicles not equipped with them. To use a stolen-vehicle retrieval system, a small transceiver first must be installed in a hidden recess of the car. If the vehicle is stolen, the transceiver emits a signal that is picked up by a tracking computer. The signal allows the manufacturer or the police to locate the stolen vehicle quickly.

Lojack Corporation, headquartered in Dedham, Massachusetts, is the only company that supplies mobile tracking computers to police departments (Figure 14.2). Other companies do their own tracking and charge a monthly monitoring fee for the service.

With a Lojack system, each transceiver is given a unique five-digit code that is paired with a vehicle identification number (VIN) in a state police crime computer. When you have installed a Lojack system and report a theft, police enter

Figure 14.2   A Lojack tracking computer installed in a police car. (*Courtesy of Lojack Corporation.*)

your car's VIN into the computer network of the state police broadcast system, which activates the Lojack transceiver in your car. Police cars equipped with Lojack tracking computers receive the broadcast signal of the transceiver and follow a homing procedure that takes them to the stolen vehicle—if it is still in the tracking area. You must report your car stolen before it is driven out of the range of the tracking computers.

The Lojack system is available in California, Florida, Georgia, Illinois, Massachusetts, Michigan, New Jersey, and Virginia. The basic model sells for about $600. Lojack System II comes with a starter disabler and sells for about $700. Lojack System III, at about $800, includes a starter disabler and passive alarm system.

Combating Carjackers

During the summer of 1991, in Cleveland, Ohio, a driver was waiting at a traffic light when four young men approached his car. After forcing him out at gunpoint, they jumped in and drove off. At about the same time, a woman in Hawthorne, California, was parking her car after returning from her lunch break. Car thieves shot and killed her and stole her car. These are two examples of a crime called *carjacking*. The term was coined in 1991 to identify a sudden rash of similar crimes occurring throughout the United States. Carjacking is a combination of car theft, assault, armed robbery—and sometimes murder. Carjacking happens in parking lots, at drive-in restaurants, at stop lights, at gas stations, and on highways.

Police records show that 416 carjackings occurred in Houston during the first eight months of 1991—an average of 52 a month. This figure was up 26 percent from the same period in 1990. In 1991, Dallas reported nearly 80 carjackings a month. In that year, Detroit faced nearly 300 carjackings a month. The crime was so rampant that the police formed a special carjacking task force. Several owners of downtown restaurants were forced into bankruptcy because many of their former customers became afraid to drive downtown. Other cities that have been touched by the carjacking epidemic include New York, Chicago, St. Louis, New Orleans, Los Angeles, San Diego, San Francisco, and Portland.

There are many theories about why carjackings have increased suddenly. Some security professionals believe that it's a response to sophisticated vehicle antitheft devices: Car thieves who aren't smart enough to defeat the protection systems turn to carjacking. The flaws in this theory are that many carjackings involve cars that have no security systems, and many of the stolen vehicles aren't sold or stripped.

Carjackers seem to have little in common with people who steal cars for money. Often, a carjacker is a wild-eyed hoodlum on the run and in need of a getaway car. Drug tests indicate that as many as 90 percent of carjackers may be high on drugs. Because carjackers are often irrational and impulsive, they work in many different and unpredictable ways. They work alone and in groups. They approach males and females of all ages. They target all types of cars at all times of day and night. It's hard to describe a "typical" carjacking.

Nevertheless, there are some common factors. By being aware of them will allow you to better protect yourself. Consider the following carjacking incidents:

- *Farmington Hills, Michigan.* While driving, a man was bumped from behind. He stopped to survey the damage. The other driver approached with a gun, ordered him out of the car, and stole it—leaving behind a car that had been stolen earlier.

- *Houston.* While shopping in an auto parts store, a 22-year-old man was shot by an armed robber. The robber fled in the man's 1990 Pontiac.

- *Philadelphia.* While at a gas station, a couple lost their BMW to a man who threatened them with a gun in a rolled-up newspaper.

- *Portland, Oregon.* A county circuit court judge had his $30,000 sports car stolen at gunpoint while he was in a supermarket parking lot.

At first, those examples may seem very different from one another, but they, like most carjackings, have two things in common: The perpetrators are most often young males (often gang members or drug addicts), and the victims are caught by surprise.

Carjackers like to target a person who seems vulnerable—someone who's day-dreaming or someone who acts friendly or trusting toward strangers. Aside from keeping your car doors locked and your windows rolled up, the most important thing you can do to protect yourself is to stay alert—not just when you're driving or sitting at a stop light but also whenever you're walking to or from your car.

When walking to your car, have your keys ready so that you can open the door quickly (instead of having to fumble around). Be aware of anyone who may be approaching you. Before getting into your car, be sure no one is crouched in the back.

When you're driving, don't automatically stop if another car bumps yours. First notice who is in the other car. Ask yourself, "Do they look like a harmless family or like young toughs?" If they seem suspicious, turn on your emergency flashers and signal the driver to follow you. Then drive slowly to a police station. If the people in the car seem to get upset, drive faster.

Chapter

15

Home and Office Insurance

By using the information given throughout this book, you'll greatly reduce the
risk that your home will catch on fire or be broken into. However, unexpected
disasters are always possible, and natural forces such as hurricanes, floods, lightning, and tornadoes can damage or destroy your home in seconds. Whether you
rent or own your home or office, you need some kind of insurance. This chapter
explains how to understand insurance policies, how to get the most coverage for
your premium payments, and how to be sure that your insurance company will
pay you when you submit a legitimate claim.

Do you need homeowner's insurance? Ask yourself whether you would face
financial hardship if you lost everything (or nearly everything) in your home
or if a visitor got hurt on your property and sued you for major medical
expenses. For most people who have to work for a living, homeowner's insurance is a must.

Understanding Your Policy

Although some insurance companies print their policies on letterhead stationery
or introduce slight alterations of wording, most homeowner's policies are based
on the HO series of policies developed by the Insurance Services Office. The policies that are most likely to be of interest to you are HO-1, HO-2, HO-3, HO-4,
and HO-6. If you understand these policies, you should have no trouble identifying the versions that your insurance company uses.

All the HO policies cover personal injury and property damage liability that
you're found responsible for anywhere in the world, damage to your personal
property, living expenses if you can't live in your home because of a fire or other
covered peril, and loss of credit cards. The HO-1 is a basic policy that covers property against the following perils: fire or lightning, windstorm or hail, explosion,
riot or civil disturbance, damage from aircraft or vehicles, smoke damage, vandalism or malicious mischief, theft, glass breakage, and volcanic eruption.

The HO-2 policy, called the *broad form*, is similar to the HO-1 policy but adds coverage of the following perils: falling objects; freezing of plumbing, heating, or air-conditioning systems; weight of ice, snow, or sleet; sudden and accidental discharge from an artificially generated electric current; accidental discharge or overflow of water or steam from a plumbing, heating, or air-conditioning system; and sudden and accidental tearing apart, cracking, burning, or bulging of a heating, air-conditioning, or sprinkler system.

The HO-3 policy, called the *all-risk policy*, is the most popular. It covers the building (but not the contents) against every peril except those it specifically excludes. Three specific exclusions are mine subsidence, earthquakes, and floods. The HO-3 covers the contents of a home only on a limited basis.

The HO-4 is a renter's insurance policy, and the HO-6 is used for condos and co-ops. Neither policy provides coverage for the building, but both cover personal property.

How Much Building Protection Is Enough?

In addition to choosing the right coverage, you need to make sure that you have designated the right dollar amount. Generally, a home owner should have the building insured for at least 80 percent of its replacement cost—not its market value. It may not be necessary to have 100 percent replacement coverage because there's little chance that your entire house will be lost. Even in a major fire, part of your house may be saved—and the insurance company won't pay you for that part. If you don't have coverage for at least 80 percent of the home's replacement value (building structure only, not contents), you may be severely penalized when you file a claim.

Your insurance company will pay a percentage of your damages based on the percentage of replacement cost stated in your policy. You have to maintain at least 80 percent of the replacement value of your home in order to receive 100 percent of the coverage amount stated in your policy. In other words, suppose that your home's replacement value is $100,000 and that you maintain $80,000 worth of coverage. If you were to suffer $40,000 worth of covered damages, the insurance company would pay you $40,000, or 100 percent of your loss. If you had maintained only $40,000 worth of coverage (half of 80 percent of the home's replacement value), the insurance company would pay you only $20,000, or 50 percent of the loss.

If you maintain coverage that represents 80 percent of your home's replacement value, you'll be self-insuring for the other 20 percent. Regardless of how much of a loss you suffer, the insurance company won't pay more than the amount of your policy.

Maintaining coverage for 80 percent of replacement value can be tricky because prices for materials change. You might want to have a little more coverage just to avoid a problem. It's a good idea to check the replacement value of your home periodically to keep your coverage up to date. Some insurers offer "inflation guard" policies that increase your coverage (as well as your premiums) automatically based on the inflation rate.

Personal or Home-Business Property

Your homeowner's policy probably will include a blanket amount of coverage for many personal items, but you may need additional coverage for some specific possessions. If you have an expensive piece of jewelry, for instance, your basic policy probably won't be enough. You might need to add a personal-articles *floater* to the policy.

If you run a business from your home, you'll need to let your insurance agent know. Depending on the type of business you have, you might need special coverage for liability or business equipment and supplies. If you run a business from your home without informing your insurance agent, the insurance company may refuse to pay your claims.

Special Disasters

Depending on where you live, your home may be vulnerable to special disasters that may not be covered by a standard homeowners' policy. Residents of Arkansas, California, Illinois, Indiana, Kentucky, Mississippi, Missouri, New York, South Carolina, and the New England States are in danger of earthquakes. In California, insurance premiums for earthquakes can cost up to $4 per $1000 of coverage. In other places, the rates are as low as 30 cents per $1000 of coverage. The problem with most earthquake insurance, however, is that the deductible may be as high as 15 percent.

Residents of Pennsylvania and other coal-mining areas are at special risk of mine subsidence damage when an underground mine collapses. Maps that show where coal once was mined are available in these regions. If you live near an abandoned mine, you should get coverage for mine subsidence damage. Pennsylvania provides this type of insurance for a home but not for the property inside.

Many people who face hurricanes and floods aren't fully covered for these two dangers. A standard homeowners' policy covers wind damage but not water damage. If a home is hit by a hurricane, the insurance company will pay claims only on the damage caused by the hurricane's wind and not by the water that it carried to the house. Consider getting flood insurance if you live near a large body of water, in an area with a high risk of hurricanes, or on land that has poor drainage and excessive runoff. You can obtain flood insurance through the National Flood Insurance Program (NFIP). For information, contact the Federal Emergency Management Agency (FEMA).

Available Discounts

When shopping for homeowners' insurance, be aware that you may be able to take advantage of one or more premium discounts. Different insurance companies give different discounts, so you may have to shop around to get the best price. Here are some things that discounts are given for:

- A newly built home
- Being age 55 or over
- Being a nonsmoker
- Having no prior losses
- Having an intruder alarm
- Having deadbolts on all exterior doors
- Having working smoke detectors
- Having a sprinkler system

In addition to using all applicable discounts, you may be able to save money by getting your homeowners' policy from the same company that insures your car. Another way to save money is by paying your premiums annually rather than quarterly. You also may want to consider lowering your premiums by raising your deductible. However, don't raise it so high that a loss will be too much of a financial burden.

Federal Crime Insurance Program

Since its founding in 1979, FEMA's mission has been clear: to reduce the loss of life and property and protect our nation's critical infrastructure from all types of hazards through a comprehensive risk-based emergency management program of mitigation, preparedness, response, and recovery.

The number and severity of natural disasters that occur each year demand that action be taken to reduce the threat that earthquakes, hurricanes, tornadoes, severe storms, floods, and fire impose on the nation's economy and the safety of its citizens. The 2000 budget allowed FEMA to continue providing crucial lifesaving and disaster-recovery activities and turn the nation's attention to the prevention of disaster losses. This demonstrates the federal administration's compassion for disaster victims and at the same time replaces those victims with responsible home owners, communities, institutions, and states. The 2000 budget also enabled FEMA to address the emergency requirements for the public's safety by enhancing state and local capability to respond, recover, and prevent natural and human-made disaster.

Under Project Impact, an initiative to build disaster-resistant communities, FEMA's work is resulting in the implementation of up-to-date risk-reduction technologies and practices. Project Impact uses concerted and proven methods for overcoming the impediments to mitigation implementation: locally based decision making, private-sector participation, Federal/state/community partnerships, federal leadership, public awareness and education, and celebrations of success. The effectiveness of the Project Impact approach is demonstrated by the facts: over 680 corporate and business partners, 128 active and committed communities, and support for Project Impact in all 50 states. In the seven pilot communities of Project Impact, the initial $2.7 million in federal funds have now leveraged over $24 million in nonfederal resources. In 2000, FEMA proposed that a separate and distinct fund for predisaster mitigation activities be established to support Project Impact.

In 2000, FEMA addressed a long-standing problem involving 35,000 properties—most of them built in unsafe locations before local flood hazard maps were drawn and therefore charged less than full-risk premiums—that have suffered repetitive flood losses and cost the NFIP over $200 million a year. These repetitive-loss properties have had two or more flood claims in the last 10 years. Of this group, 7300 have had four or more claims and account for approximately $60 million of the $200 million. A smaller subset of 201 homes has had 10 or more flood claims over this period. FEMA is requesting an appropriation under the National Flood Mitigation Fund, which supports flood mitigation assistance planning at the state level, to remove or elevate these repetitive-loss properties from the floodplain. The end result will be a lower net subsidy required to operate the NFIP, fewer claims on the Disaster Relief Fund, and fewer individuals living in hazardous areas.

The growing awareness of flood hazards in this country is demonstrated by the expanded use of flood-map panels from their original role of supporting flood risk determinations for the NFIP. Use of the maps now is required for every building permit issued and mortgage transacted by federally regulated financial institutions. Given the rapid pace of development and the changes in technology, program funding from the federal policy fee has not been sufficient to maintain the accuracy of the flood hazard data in developing areas and to convert the maps from a manual to a digital format. This conversion is necessary for rapid updates, Internet distribution, and automated determinations and will lead to production/reproduction efficiencies. The 2000 budget proposed a one-time appropriation for the Flood Map Modernization Fund to develop new technical standards and procedures necessary to implement the flood-map modernization plan. The budget also included a proposed mortgage transaction fee to support the multiyear flood-map modernization activities that represent a shift from taxpayer-funded disaster aid toward self-protection.

As part of the administration's solution to the real threat of terrorism, FEMA is seeking funding for antiterrorism training and exercise activities to enhance the capabilities of state and local governments and first-responders, interagency planning activities, emergency public information, protection of FEMA assets, and coordination of terrorism-related strategy and policy.

The philosophy of mitigation and preparedness as investments for the future is reflected in FEMA's dam safety and fire enhancements included in the 2000 budget request. An additional $2 million investment in FEMA's dam safety program will enhance state preparedness aimed at reducing the risk posed by unsafe dams. FEMA's request for an additional $10.4 million for its fire training, research, and education programs expands the availability of training offered to the nation's firefighters and emergency services personnel, as well as outreach programs aimed at improving public fire safety, particularly in groups at greatest risk from fire.

In 2000, FEMA implemented an Emergency Management Performance Grant (EMPG) that consolidates funding for FEMA's nondisaster programs, for which state emergency management agencies are the primary recipients, into a grant with one source of funding. The budget redistributes pass-through grant funds under emergency management planning and assistance from the preparedness and mitigation activities and proposes appropriation transfers from the Disaster Relief Fund and Pre-Disaster Mitigation Fund to the Office of Financial Management under the executive direction activity. The EMPG streamlines the application, financial, and progress-reporting processes, provide flexibility for states to target funds (with the exception of terrorism) to meet emergency management priorities, and allow for efficient use of limited financial and staff resources for both FEMA and the states.

The budget included an increase for the Emergency Food and Shelter (EFS) Program in 2000. Recent findings show an alarming decline in the availability of food in food pantries and soup kitchens. In addition, shelters and emergency service agencies are reporting a marked increase in the number of individuals, especially the elderly and unaccompanied children, and families with children requesting food and housing assistance. Many of the adults are the working poor who must choose between buying food and paying rent. Medical emergencies are also causing a decrease in funds available for food and shelter. The 2000 request enabled the EF S Program to provide additional meals and nights of shelter while increasing payments for rent, mortgages, and utility bills for those in need.

In order to carry out FEMA's mission, provide the services the public has come to expect from the agency, and support the initiatives just described, it is necessary for FEMA to have an adequate, well-trained workforce. Over the years, the buying power of FEMA's salaries and expenses appropriations has been eroded by the need to absorb increases in the cost of doing business. For example, funding unforeseen items such as the unplanned increase in the 1999 pay raise, additional rent, and moving costs and workers' compensation costs, among others, required reductions in the number of work years that the agency could afford in 1999. The 2000 request sought funds to restore the dollars lost to unbudgeted increases in 1999 and to pay for uncontrollable increases in pay and space rental costs for 2000. Providing adequate training funds for its workforce was another priority for 2000, with a requested increase of $883,000. Part of this workforce training effort involved an exchange program whereby regional personnel received experience working at headquarters and vice versa. In addition, an increase in funds for permanent change of station moves allowed more individuals to be selected for positions outside their permanent-duty stations. An increase of $2.9 million supported the planning, training, and interagency coordination necessary for the nation's emergency managers to prepare for potential terrorism incidents. In order to further promote and support FEMA's already highly successful effort to make communities disaster resistant, the 2000 budget requested $2.5 million for administrative expenses associated with Project Impact. For its initiative to expand its fire programs, FEMA requested $1 million. All these enhancements to FEMA's salaries and expenses appropriation allowed FEMA to lead the nation's emergency managers into the next century.

Highlights

The major highlights of each budget activity follow:

Response and recovery

This activity focuses on providing FEMA services to communities stricken by disaster with an increase in timeliness, refining program delivery activities to effect increased cost efficiency, and increasing internal and external customer satisfaction with the delivery of services. The request of $58,349,000 for response and recovery included resources to continue to enhance preparedness measures to address terrorist incidents at the federal and regional levels. An increase of $4.2 million, coupled with the 1999 congressionally directed increase of $.8 million, allowed FEMA to continue upgrading and/or replacing the equipment in its Mobile Emergency Response System (MERS).

Preparedness

The 2000 budget included $34,800,000 for this activity, through which FEMA attempted to build and maintain a collaborative framework of federal, state, local, and private-sector resources to yield a general reduction in the risk of loss of life and property from all hazards. The focus of this strategy is on emergency management professional development; on the establishment of capability standards and assessment through tests, exercises, and real world experiences, including the threat of terrorism; on supporting planning and public education; and on creating partnerships with the private sector and other nations. Beginning in 2000, grants to states previously included in preparedness were consolidated into the EMP Grant program.

Fire prevention and training

The goal of the U.S. Fire Administration is to reduce the risk of loss of life and property from fire and fire-related hazards. The request of $45,130,000 included resources for increased development and delivery of training and training materials, including support for antiterrorism training, for fire and emergency services personnel, as well as critical research to address the nation's fire problem. Increases also were requested to address fire incident reporting, support outreach initiatives, and provide for required repair projects at the National Emergency Training Center facility. These increases supported recommendations made by a blue ribbon panel convened by the director to review FEMA's fire programs.

Operations support

In addition to providing agency-wide services such as logistics support for day-to-day, emergency, and disaster operational needs, the programs administered under this budget activity were designed to provide a healthful, safe, and secure environment for FEMA personnel, facilities, and equipment and its emergency management partners. The request of $32,858,000 provided for enhanced protection of agency facilities and equipment and continuation of a multiyear health and safety upgrade project at federal regional centers in Region I (Maynard, MA), Region IV (Thomasville, GA), and FEMA headquarters that will bring these facilities into compliance with Occupational Safety and Health Administration (OSHA) regulations and standards, as well as provisions of the Americans with Disabilities Act.

Information technology services

FEMA requested $44,090,000 for information technology services in 2000. This activity provided leadership and management of information technology resources for the agency. An increase of $1,500,000 initiated a 3-year initiative to integrate FEMA's voice and data network components in order to streamline operations and reduce costly maintenance.

Mitigation programs

This activity provided for the development, coordination, and implementation of FEMA's mitigation strategy. FEMA's Project Impact initiative is part of this strategy and is designed to encourage communities and their citizens to take informed and effective mitigation actions before a disaster in order to reduce the long-term risk to people and property from hazards and their effects. The 2000 request included $2 million for dam safety training, research and assistance to states, bringing the total for the National Dam Safety Program to $5.5 million for 2000, as authorized by the National Dam Safety Program Act. The request also included $1 million to determine the feasibility of developing a mitigation strategy to protect the considerable federal research investment in universities that are located in high-hazard areas. Of the total request of $126,391,000, $96,760,000 would be charged directly either to the National Flood Insurance Fund or the National Flood Mitigation Fund. An increase of $2,893,000 in NFIP funds would allow FEMA to initiate and complete more flood restudies and map updates for more accurate flood hazard identification. In order to strengthen the ideology of addressing specific risks faced by a particular community, the Performance Partnership Agreement grants provided for the consolidation of emergency management planning and assistance appropriation in 2000 with other emergency management grants.

Policy and regional operations

This activity provided direction to FEMA's activities related to strategic planning, regional policy coordination, and intergovernmental affairs. The 2000 request of $15,203,000 included $1,564,000 for projects designed to improve the delivery of FEMA's programs at the regional level.

Executive direction

In 2000, the request for this activity totaled $186,165,000, of which $156,000 was charged to the NFIF. The request included a total of $141,951,000 for FEMA's Consolidated Emergency Management Performance Grant Program. This grant program, which included transfers from the Preparedness, Mitigation, Disaster Relief Fund, and Pre-Disaster Mitigation Fund budget activities, consolidated funding for existing FEMA nondisaster programs, for which state emergency management agencies are the primary recipients into a grant with one source of funding, streamlined the application and financial and progress reporting processes, provided flexibility for states to target funds (with the exception of terrorism) to meet emergency management priorities, and allowed for efficient use of limited financial and staff resources for both FEMA and the states. The total funding level proposed for the EMPG represented a $12.450 million increase in funding assistance to states when compared with equivalent funding sources included in the 1999 Performance Partnership Agreement cooperative agreements. Included in the 2000 request was an increase of $8,300,000 for state and local improvements to plans for response to terrorist incidents.

FEMA requested an increase of $926,000 for executive direction to promote technical understanding of terrorism-related policy issues and develop methods to provide instructions to the public in the event of a terrorist incident. An additional $800,000 was requested to address critical infrastructure requirements under FEMA's national security programs.

Disaster Relief Fund

For 2000, FEMA requested an appropriation of $300 million for the Disaster Relief Fund (DRF), which included $126,957,000 for disaster support costs and $2,900,000 to be transferred to the emergency management planning and assistance appropriation into the Consolidated Emergency Management Performance Grant Program for disaster preparedness improvement grants. In addition, FEMA requested an emergency contingency appropriation of $2,480,425,000. The combined request represented the five-year average obligations for disaster-specific costs, less Northridge, plus disaster-support costs.

Pre-Disaster Mitigation Fund

Under this initiative, FEMA provides funds for community-identified mitigation projects that reduce the exposure to disaster losses. In 2000, FEMA requested $30 million for the Pre-Disaster Mitigation Fund. As in the past, these funds were expected to leverage private-sector resources. Of the total amount requested, $2.6 million was transferred to emergency management planning and assistance appropriation into the Consolidated Emergency Management Performance Grants to support Project Impact mitigation activities at the state level.

National Flood Insurance Fund

The National Flood Insurance Program makes flood insurance available in communities that adopt measures to reduce losses from future flooding. It also provides funding for floodplain management and flood mitigation at the state and local levels. As in previous years, the fund continued to pay all costs associated with floodplain management and flood mitigation assistance planning support to states and the salaries and expenses of staff assigned to flood mitigation and flood insurance operations from the $30 fee collected from policyholders.

National Flood Mitigation Fund

Through fee-generated funds transferred from the National Flood Insurance Fund, this fund supports activities to eliminate preexisting, at-risk structures that are flooded repetitively and provides flood mitigation assistance planning support to states. In 2000, an appropriation of $12 million was requested to increase the number of buyouts of structures that suffer repetitive losses. The end result will be a lower net subsidy required to operate the NFIP, less claims on the Disaster Relief Fund, and fewer individuals living in hazardous areas.

Flood-Map Modernization Fund

FEMA proposed this fund in 2000 to support the use of state-of-the-art technology to improve the accuracy and completeness of flood hazard information, make the information more readily available, and continue to educate the public regarding the risks of flood hazards. A one-time appropriation of $5 million was requested in 2000. In addition, FEMA proposed to assess and collect fees from mortgage transactions to support flood-map modernization activities.

Emergency food and shelter

This appropriation provided grants to voluntary organizations at the local level to supplement their programs for emergency food and shelter. FEMA requested an increase of $25 million for emergency food and shelter in 2000. This increase enabled the Emergency Food and Shelter Program to provide approximately 6,653,968 additional meals and 964,309 additional nights of shelter. Further, 44,761 more rents and mortgages and 44,666 additional utility bills were paid.

Inspector General

FEMA requested $8,015,000 for this appropriation. An increase of $2,615,000 would allow the Office of Inspector General to fulfill its statutory mandate of conducting independent and objective audits and investigations, including those for Project Impact, counter- and antiterrorism, and the national mitigation strategy.

National Insurance Development Fund

The budget request for 2000 reflected the national performance review (NPR) recommendations to eliminate the FCIP as a federally sponsored government program and devolve to the states and private sector. The Federal Crime Insurance Program authorization expired September 30, 1995. For 2000, FEMA requested forgiveness for all prior borrowings of the program and that outstanding interest on these borrowing be canceled.

16

Home and Office Security Basics

Throughout this book I've covered lots of information about safety and security systems, devices, and hardware for homes and offices. I've discussed the importance of getting homeowner's insurance and of using psychological self-defense. If you've read all the other chapters, you know more about home security than most burglars do, and you have all the information necessary to think like a security professional.

This chapter discusses how you can put everything together to make your home or office as safe as you want it to be. It's important to understand that no single security plan is best for everyone. Each home and neighborhood has unique strengths and vulnerabilities, and each household has different needs and limitations.

The important limitation most of us face is money. If money were no object, it would be easy for me to lay out a great security plan for you. I would advise you to build a steel-frame-construction Smart House with steel doors, multiple-bolt locks, high-security cylinders, bullet-resistant glass, and an integrated CCTV system. I also might advise you to hire armed guards. Such investments would make your home or office more secure, but they would be overkill.

With proper planning, you can be safe in your home without spending more money than you can afford or being too inconvenienced. Proper planning is based on the following considerations:

How much money are you willing to spend?

How much risk is acceptable to you?

How much inconvenience is acceptable to you?

How much time are you willing to spend on making your home more secure?

How much of the work are you willing to do yourself?

No one can answer these questions for you. You are the best person to design a security plan that meets your needs.

Before you can create the best plan for your home or office, you'll need to conduct a safety and security survey (or vulnerability analysis). The survey requires your walking around the outside of your home or office and through every room. You should make note of all potential problems.

Surveying a House or Office

The purposes of a safety and security survey are

- To help you to identify potential problems

- To assess how likely and how critical each risk is

- To determine cost-effective ways to either eliminate the risks or bring them to an acceptable level

The survey will allow you to take precise and integrated security and safety measures. A thorough survey involves not only inspecting the inside and outside of your home or office but also examining all your safety and security equipment and reviewing the safety and security procedures used by all members of your household. The actions people take (or fail to take) are just as important as the equipment you may buy. What good are high-security deadbolts, for instance, if residents often leave the doors unlocked?

As you conduct your survey, keep the information in the preceding chapters in mind. You'll notice many potential safety and security risks (every home and office has some). Some of the risks will be simple to reduce or eliminate immediately. For others, you'll need to compare the risk to the cost of properly dealing with them. There's no mathematical formula to fall back on. You'll need to make subjective decisions based on what you know about your household and neighborhood and guided by the information provided throughout this book.

When surveying your house or office, it's best to start outside. Walk around the building and stand at the vantage points that passersby are likely to have. Many burglars will target a home because it's especially noticeable while driving or walking past it. When you look at your home from the street, note any feature that might make someone think that no one is home, that a lot of valuables might be in the home, or that the home might be easy to break into.

Keep in mind that burglars prefer to work in secrecy. They like heavy shrubbery or large trees that block or crowd an entrance, and they like homes that aren't well lighted at night. Other things that may attract burglars' attention include expensive items that can be seen through windows, a ladder near the home, and notes tacked onto the doors.

As you walk around the home, note anything that might help to discourage burglars. Can your "Beware of Dog" sign or your fake security system sticker be seen in the window? Walk to each entrance and consider what burglars might like and dislike about it. Is the entrance well lighted? Can neighbors see someone who's at the entrance? Is there a video camera pointing at the entrance? Does the window or door appear to be easy to break into?

After surveying the outside of your home, go inside and carefully examine each exterior door, window, and other opening. Consider whether each one is secure but allows you to get out quickly. Check for the presence of fire safety devices. Do you have enough smoke detectors and fire extinguishers? Are they in working order? Are they in the best locations?

Take an honest look at the safety and security measures that you and your family take. What habits or practices might you want to change?

Home Safety and Security Checklist

Because every family, home, and neighborhood is unique, no safety and security survey checklist can be comprehensive enough to cover all of every home or office's important factors, but the following checklist will help to guide you during your survey. Keep a notepad handy to write down details or remedies for potential problems.

As you conduct your survey, note each potential problem that is of concern to you.

Home exterior

Shrubbery. (Shouldn't be high enough for a burglar to hide behind or too near windows or doors.)

Trees. (Shouldn't be positioned so that a burglar can use them to climb into a window.)

House numbers. (Should be clearly visible from the street.)

Entrance visibility. (Should allow all entrances to be seen clearly from the street or other public area.)

Lighting near garage and other parking areas. (Should allow clear view of anyone present.)

Ladders. (Shouldn't be in the yard or in clear view.)

"Alarm System" or "Surveillance System" stickers. (Shouldn't identify the type of system that's installed.)

Mailbox. (Should be locked or otherwise adequately secured and should show no name or only a first initial and last name.)

Windows. (Should be secured against being forced open but should allow for easy emergency escape.)

Window air conditioners. (Should be bolted down or otherwise protected from removal.)

Fire escapes. (Should allow for easy emergency escape but not allow unauthorized entry.)

Exterior doors and locks

Included here are doors connecting a garage to the home.

Door material. (Should be solid hardwood, fiberglass, PVC plastic, or metal.)

Door frames. (Should allow doors to fit snugly.)

Door glazing. (Shouldn't allow someone to gain entry by breaking it and reaching in.)

Door viewer (without glazing). (Should have a wide-angle door viewer or other device to see visitors.)

Hinges. (Should be either on inside of door or protected from outside removal.)

Stop molding. (Should be one piece or protected from removal.)

Deadbolts. (Should be single cylinder with free-spinning cylinder guard and a bolt with a 1-inch throw and a hardened insert.)

Strike plates. (Should be strike boxes or securely fastened.)

Door openings (mail slots, pet entrances, and other access areas). (Shouldn't allow a person to gain entry by reaching through them.)

Sliding-glass doors. (Should have a movable panel mounted on interior side and a bar or other obstruction in the track.)

Inside the home

Fire extinguishers. (Should be in working order and mounted in easily accessible locations.)

Smoke detectors. (Should be in working order and installed on every level of the home.)

Rope ladders. (Should be easily accessible to bedrooms located above the ground floor.)

Flashlights. (Should be in working order and readily accessible.)

First-aid kit. (Should contain fresh bandages, wound dressing and burn ointments, aspirin, and plastic gloves.)

Telephones. (Should be programmed to dial the police and fire departments quickly, or their phone numbers should be posted nearby.)

Burglar alarm. (Should be in good working order and adequately protected from vandalism and should have adequate backup power.)

Safes. (Should be installed so that they can't be seen by visitors.)

People in the home

Doors (locking). (Should be locked by all residents every time they leave the home—even if they plan to be gone for only a few minutes.)

Doors (opening). (Should not be opened by any resident unless the person seeking entry has been satisfactorily identified.)

Fire escape plan. (Should be familiar to all residents as a result of practicing how to react during a fire.)

Confidentiality. (Should be maintained by all residents regarding locations of safes, burglar alarms, and other security devices.)

Money and valuables. (Should be stored in the home only in small amounts and only if they have low resale or "fencing" value.)

Drapes and curtains. (Should be routinely closed each night.)

Garage doors. (Should be kept closed and locked.)

When you've finished reading this book, keep this checklist handy. If you are intent on managing your home or office security to your best advantage, you'll review the checklist periodically to be sure that you've responded to your changing security needs.

Surveying a Building

In many ways, surveying an apartment or office is like surveying a house. The difference is that you have to be concerned not only about the actions of your household but also about those of your landlord, the apartment managers, and the other tenants. The less security conscious others around you are, the more at risk you will be. No matter how much you do to avoid causing a fire, for example, a careless neighbor might cause one. If your neighbors don't care about crime prevention, your apartment building or complex will be more attractive to burglars.

As you walk around the outside of your apartment or office, notice everything that would-be burglars might notice. Will they see tenants' "Crime Watch" signs? Will they see that all the apartments have door viewers and deadbolt locks? Burglars hate a lot of door viewers because they never know when someone might be watching them.

After surveying the outside, walk through your apartment and look at each door, window, and other opening. If you notice major safety or security problems, point them out to your landlord. You also might want to suggest little things that the landlord can do to make your apartment or office more secure.

High-Rise Apartment Security

High-rise apartments have special security concerns that do not apply to apartments with fewer floors. In a high-rise, more people have keys to the building, which means that more people can carelessly allow unauthorized persons to enter.

The physical structure of a high rise often provides many places for criminals to lie in wait for victims or to break into apartments unnoticed. Many high-rise buildings aren't designed to allow people to escape quickly during a fire.

The safest apartments have only one entrance for tenants to use, and that entrance is guarded 24 hours a day by a doorman. An apartment that doesn't have a doorman should have a video intercom system outside the building. Video intercoms are better than audio intercoms because they let you see and hear who's at the door before buzzing the person in.

Neighborhood Crime Watches

A neighborhood crime watch is one of the most effective and least costly ways to prevent crime and reduce fear. It bonds area residents to help reduce crime and improves relations between police and the communities they serve.

Neighborhood crime watches encourage citizen participation in reducing crime. Residents and police work together to achieve the common goal of making their neighborhoods safer. Basically, members of a neighborhood watch meet their neighbors and learn how to make their homes more secure and how to look out for and report suspicious activities to police. Neighborhood crime watches also help a community to build pride and serve as springboards for efforts that address other community concerns. It's harder for criminals to get away with crimes in areas where neighbors have banded together to watch out for criminal activities.

Basic Activities of a Neighborhood Crime Watch

Meetings. These should be held at least once a month.

Host speakers. These include as police officers, locksmiths, intruder alarm installers, security consultants, and others who can provide useful home and neighborhood security information.

Citizens' patrol. This consists of crime watch members who walk or drive through the neighborhood and alert police to crime and questionable activities. (Not all neighborhood crime watches need a citizens' patrol.)

Communications. This includes a newsletter that provides information about the neighborhood crime watch and helpful security tips.

Special events. These include block parties, various social events, and crime-prevention programs.

The Alachua County Sheriff's office states:

Neighborhood Crime Watch

The Alachua County Sheriff's Office recognizes that the responsibility for making your home and family safe from crime extends beyond routine Sheriff's Office patrol. This is why we believe in Neighborhood Crime Watch, a cooperative effort between the citizens of Alachua County and the Alachua County Sheriff's Office. Residential burglary is a major crime that can be effectively combated with neighborhood awareness. It is difficult for law enforcement to effectively cover all residential areas. It is also difficult for law enforcement to completely familiarize themselves with all the persons, vehicles and surroundings in a area. Concerned citizens can become the eyes and ears of their Sheriff's Office by immediately reporting suspicious persons, vehicles, and activities. The duties of the citizen does not include apprehension of any suspects. Apprehension is still the responsibility of law enforcement. Remember, an informed, alert, and active community is absolutely necessary to win the war on crime.

What Is Neighborhood Crime Watch?

Neighborhood crime watch is the involvement of citizens in cooperation with their local law enforcement agency to reduce criminal activity in the neighborhood. It consists of residents within the neighborhood who recognize any suspicious criminal or dangerous activity and report it immediately to the local law enforcement agency.

What Are Suspicious Activities to Report?

You may become aware of something that is out of the ordinary that you feel is leading up to some type of criminal activity. This is your chance to prevent a crime in your neighborhood. Do not be afraid to call the police when you hear or observe any of the following:

- Screams or sounds of distress
- Loud or unusual noises
- Strange persons loitering around a neighbor's house while the neighbor is away
- Strange vehicles that are parked or driving in the neighborhood with no apparent reason
- A stranger carrying electronic goods, household items, or similar devices
- Broken or open doors or windows
- Persons walking down the street repeatedly peering into parked cars

---

Why Neighborhood Crime Watch Works

Most law enforcement agencies have limited staffing. Your cooperation as a resident will help greatly to overcome this shortage by the watchful eyes and ears of the citizens observing suspicious activities within their own neighborhoods.

Citizens provide the benefit of having someone watching their neighborhood at all times. This lets criminals know that they are being observed.

Operation Crime Watch in Plano, Texas:

> Neighborhood Crime Watch operates under two principles: know your Neighbors and Report Suspicious Activity.
>
> The basic idea is for neighbors to watch out for each other. By looking after a neighbor's property as if it were your own, you will more likely contact the police if you observe something suspicious. Citizens act as extra "eyes and ears" for the police department so that calls to the police will be made when criminal activity is suspected. An alert and cooperative neighborhood is the greatest single defense against crime. By getting to know your neighbors and their vehicles, you will most likely be alert to suspicious people, vehicles, and/or sounds that could indicate criminal activity.
>
> The Crime Prevention Unit of the Plano Police Department will help you organize a Crime Watch Area. Once the Area is organized, Officers will send bulletins to Crime Watch Areas on an as-needed basis. This information is sent regarding crimes of a serious nature, or when a suspects description is available. This information, coupled with the two basic principles of Crime Watch, can help citizens become a part of reducing the overall crime rate in their neighborhoods.

Operation Crime Watch in Baltimore, Maryland:

> The mission of Baltimore Maryland's Operation Crime Watch is to prevent and reduce crime in Baltimore by creating and supporting neighborhood-based block watch and citizen patrol programs through a partnership between the citizens of Baltimore, the Baltimore Police Department, the Mayor's Office of Neighborhoods and the Washington/Baltimore High Intensity Drug Trafficking Area. Citizens and police officers support each other in preventing and reducing crime through more effective communication and by creating problem-solving relationships appropriate to each community.
>
> Operation Crime Watch supports and enhances block watch and citizen patrol efforts currently underway. These also serve as the models for establishing new efforts throughout Baltimore. Priority for new programs is given to targeted neighborhoods, including those designated as Strategic Neighborhood Action Plan (SNAP) areas and those receiving city-funded Hot Spot monies.

Key Partners

New Operation Crime Watch programs are created through the collaborative efforts of

> *Trained citizens* who serve as the block watchers and citizen patrollers,
>
> *Community associations* and other neighborhood-based organization such as a religious institution, PTA or block club that will sponsor the local Crime Watch effort by recruiting volunteers and hosting trainings,
>
> *Baltimore Police Department districts* that help identify the appropriate type of crime watch activity for a particular neighborhood and in some instances serves as the direct link to the trained citizen block watchers, and
>
> *Mayor's Office of Neighborhoods* provides staffing and overall coordination of Operation Crime Watch.

Three Models for Crime Watch Programs

Block watch

Citizens are given a block watch number and training to bring about more effective and more numerous calls to 911.

Block watch with block captains

Where appropriate, block watch programs include block watch captains who receive information from the trained block watchers that is forwarded on a regular basis to designated officers in police districts.

Citizen patrols

Citizens trained as block watchers participate in regular patrols of designated areas. These patrols may be on foot or in cars and usually will be in radio contact with an assigned police car to respond as needed.

How to Start a Neighborhood Crime Watch

Typically, a neighborhood crime watch gets started when a neighborhood faces a rash of crimes, such as burglaries, car break-ins, rapes, muggings, or worse. Only two or three people are needed to start a crime watch. The first step is to define the neighborhood on which the group will focus. Then schedule a date, place, and time to hold the first crime watch meeting. Go to each resident within your neighborhood boundaries to survey their concerns about crime and to ask them to attend the first meeting. Two people should go to each home, and one should carry a clipboard to keep notes about crime-related concerns. If you can afford to do so, create a flyer to give to each resident. The flyer should state the purpose of the neighborhood crime watch and the date and time of the first meeting and include a phone number to call for more information.

Contact your local police and ask if a police officer can attend your first meeting. Many police departments have a crime-prevention officer who will help the group to get started on the right foot. The City of Plano, Texas, offers the following suggestions:

- Don't worry if only a few people attend the first meeting. More people will join as they see the results of the crime watch. Ask people to come 15 to 30 minutes early so that they can meet their neighbors. Always start meetings on time, and end them on time. If the meetings run too long, people will stop showing up.

- Have a clear agenda for each meeting, and start the meetings with the most important things first. At the first meeting, have the group come up with a name. The group's name should be easy to remember, not be too long, and specify the neighborhood. An example of a good name is the "Southeast Erie Crime Watch." Also at the first meeting, elect one or more block captains and a secretary.

Block captain responsibilities

1. Pass on the information received from the area coordinator to the individual residential households.

2. Meet with residents on a semiannual basis (a crime-prevention officer can make presentations at these meetings or you might have a block party).

3. Distribute home security manuals to each of the residents and collect money for the neighborhood crime watch signs.

4. Maintain, update and distribute new neighbor packets. New neighbor packets contains a cover letter and neighborhood phone and/or e-mail lists, nine-house sheet, various pamphlets, and a McGruff house application.

5. Notify the area coordinator if block captains are no longer able to perform their duties.

18

Computer Security

Whether your computer runs Microsoft Windows, Apple's Mac OS, LINUX, or something else, the security issues are similar and will remain so even as new versions of your system are released. The key is to understand the security-related problems that you need to think about and solve.

Your home or office computer is a popular target for computer hackers because they want the information you've stored there. Hackers and crackers look for passwords, credit-card numbers, bank account information, and anything else they can find. By stealing that information, they can use your money to buy themselves goods and services.

It isn't just money-related information they're after. Hackers also want to use your computer's hard disk space, your processor, and your Internet connection. The intruder uses those resources to attack other computers on the Internet. The more computers an intruder uses, the harder it is for law enforcement to figure out where the attack is really coming from. The intruder must be found before he or she can be stopped and prosecuted.

Intruders pay special attention to home and office computers because such computers aren't very secure and are easy to break into. Because the computers often use high-speed Internet connections that are always turned on, intruders can quickly find and then attack home or office computers. They also attack computers that are connected to the Internet through dial-up connections, but hackers favor computers connected with high-speed cable and DSL modems. Regardless of how a computer is connected to the Internet, a computer is susceptible to hacker attacks.

Firewalls

A key to Internet security begins with a properly configured Internet *firewall*—which is a piece of software or hardware that helps to screen out hackers, viruses, and worms that try to reach your computer over the Internet. (Both

Windows XP Home Edition and Windows XP Professional with Service Pack 2 have a firewall already built in and active.) If you have Microsoft Windows XP (SP2) running on your computer, you can check to see if your firewall is turned on through the Windows Security System. Just click "Start," and then click "Control Panel." Click "Security Center," and then click "Windows Firewall." Versions of Windows before Windows XP did not come with a built-in firewall. If you have a different version of Windows, such as Windows 2000, Windows Millennium Edition (ME), or Windows 98, get a hardware or software firewall from another company and install it.

If you use Windows XP but want different features in a firewall, you can use a hardware firewall or a software firewall from another company. Many wireless access points and broadband routers for home networking have built-in hardware firewalls that provide good protection for most home and small-office networks. Software firewalls are a good choice for single computers, and they work well with Windows 98, Windows ME, and Windows 2000.

If you aren't sure of which version of Windows you have, click "Start," and then click "Run." In the "Run" dialog box, type "winver," and then click "OK." The dialog box that appears will tell you which version of Windows software you are running.

If you have two or more computers in your home or office network, you need to protect each computer in the network. Enabling the Internet connection firewall on each connection will help to prevent the spread of a virus from one computer to the other(s) in your network if one of your computers becomes infected with a virus. If a virus is attached to an e-mail message, however, the firewall won't block it, and it can infect your computer.

You only need to run one firewall per computer. Running multiple software firewalls isn't necessary for typical home computers, home networking, and small-business networking scenarios. Using two firewalls on the same connection could cause problems with connectivity to the Internet. One firewall, whether it is the Windows XP Internet Connection Firewall or a different software firewall, can provide enough protection for your computer.

Antispyware

Spyware programs are small applications that can get installed on your computer without your knowledge. Such programs can get installed either by downloading innocent-looking software programs that include them or through ActiveX controls hidden within the source code of participating Web sites or pop-up advertisements while you're surfing the Internet. These bundled programs and ActiveX controls can install a wide range of unwanted software onto a user's computer.

In addition to doing a detailed check of your browser history, they install a wide assortment of Dynamic Link Libraries (DLLs) and other executables files. They send a continuous data stream to the parent marketing company out from your computer and leave a backdoor open for hackers to either intercept your personal data or enter your computer. Spyware programs can install other programs directly onto your computer without your knowledge, they can send and receive "cookies" to or from other spyware programs and invite them into your computer (even if you have cookies disabled), and they can bring Trojan horses into your system that perform a wide range of "mischief" on your system—including changing your home page and downloading unwanted images and information.

Many spyware programs are independent executable files that are self-sufficient programs that take on the authorization abilities of the user. They include automatic install and update capabilities and can report on any attempts to remove or modify them. These programs can hijack your home page; reset your browser favorites; reset your auto signature; disable or bypass your uninstall features; monitor your keystrokes on- or offline; scan files on your hard drive; access your word processor, e-mail, and chat programs; and change home pages. Many spyware programs can read, write, and delete files and, in some instances, even reformat your hard drive. And they do these things while sending a steady stream of information back to the advertising and marketing companies.

Most of these programs cannot be deleted from your system by normal methods and leave residual components hidden on your system to continue monitoring your online behavior and trying to reinstall themselves. Many people notice a big decrease in their computers performance after installing spyware-infested programs—which use up your system resources.

New types of spyware are becoming more malicious: CoolWebSearch makes browsers useless by changing Internet Explorer settings and installing malicious applications; KeenValue collects information about users and sends advertisements to their systems; Perfect Keylogger logs keystrokes users enter, putting users' personal information at risk; and Marketscore redirects traffic from a host system to another that collects data before traffic reaches its final destination. Windows users should install and run an antispyware program such as Microsoft AntiSpyware, Pest Control, Spyware Doctor, Spy Sweeper, and Spybot. AOL users with current software have built-in spyware protection.

Operating System Safeguards

Many operating systems have good security features built into them, but such features are worthless if you don't know about them. One way hackers and crackers break into computers is by sending an e-mail with a virus in it. When you read the e-mail, the virus gets activated, creating an opening that the intruder can use to access your computer. Other times a hacker will take advantage of a flaw or weakness in one of your programs to gain access. When a virus gets into your computer, it may install new programs that let a hacker continue to use your computer even after you plug the holes that were used to plant the virus in the first place. Such a "back door" often is cleverly disguised so that you won't recognize it.

Securing Your Web Browser

Web browsers such as Internet Explorer, Mozilla Firefox, and Safari, as examples, are installed on most home and office computers. Because Web browsers are used so often, it's important to configure them securely. Often the default settings for Web browsers aren't set securely. If they aren't configured securely, hackers easily can gain control over your computer.

Compromised or malicious Web sites often take advantage of vulnerabilities in Web browsers either because the Web browser's default settings are set to increase functionality rather than provide security or because new vulnerabilities are discovered after the Web browser has been packaged and distributed by the manufacturer. It's important to know which features of your browser make it less secure. The Active X software feature, for example, has a long history of vulnerabilities. ActiveX is used on Microsoft Internet Explorer and allows applications or parts of applications to be used by the Web browser. It allows for extra functionality when Web browsing but also creates security problems. Cross-site Scripting (CSS or XSC) also creates vulnerabilities.

To increase your security, go to "Internet Options," and set the security level to "high." This is the safest way to browse the Internet, but you'll have less functionality and may not be able to use some Web sites. If you have trouble browsing a certain Web site, send an e-mail to the Webmaster and ask him or her to design the site so that you can browse it more securely.

If you want to browse a Web site that you trust, set the security of your "Trusted Site" to "medium." Then, when you access a Web site that you trust doesn't have malicious code, you will be able to use ActiveX and Active Scripting. This will allow you to stay safe from most Web sites and have full functionality on the sites you trust.

If you're using Internet Explorer, the privacy tab contains settings for "cookies," which are small text files placed on your computer to keep tabs on you. A cookie can contain any information that the Web site wants to have. Most cookies are harmless ways for merchants to know which Web pages you've viewed, your preferences, and your credentials. If you're using Internet Explorer, set your advanced privacy setting to "prompt" for "first party" and "third party" cookies. This will allow you to decide if you want cookies from a site. For your convenience, you can select the "Sites" button and use the "Per Site Privacy Action" option to automatically accept or reject cookies from specific sites.

Many Web browsers allow you to store passwords. For maximum security, though, it's better not to use that feature. If you do use that feature, at the "Privacy" category, go to the subcategory "Passwords" and set a master password to encrypt the data on your computer. This is especially important if you use Mozilla Firefox to manage your passwords. Next, go to the "Advanced JavaScript Settings" and disable all the options displayed in the dialogue.

Mozilla Firefox

Mozilla Firefox has many of the same features as Internet Explorer, except the "ActiveX" and the "Security Zone" model. To edit the features of Mozilla Firefox, select "Tools" and then "Options." Under the "General" category, you can set Mozilla Firefox as your default Web browser. Under the "Privacy" category, select the "Passwords" subcategory to manage stored passwords. Choose a master password.

Apple Computer's Safari

The Safari contains many of the same features and weaknesses of Mozilla Firefox. To change the Safari settings, click on "Safari" and then "Preferences." Under the Safari menu, you can choose to block pop-ups. Blocking pop-ups will make your computer more secure but may cause you to lose functionality at many sites that use pop-ups to give relevant information.

Other Security Measures

Use an antivirus program. Set it to check your e-mail before you open it. Enable the program to get updated definitions automatically. This will let you keep the antivirus program as up-to-date as possible. Also set your other software programs to receive updates automatically, if possible. When a vendor learns of a vulnerability in its software, it creates patches to make the program more secure.

In many cases, a computer will have multiple Web browsers. Even if you only use one, it's important that you configure each of them for maximum security.

Installing and Using an Antivirus Program Checklist

To make sure that you have an effective antivirus program, ask the following questions:

1. Do you have an antivirus program installed?
2. Do you check frequently for viruses before sending or receiving e-mail?
3. Do you check for new virus signatures daily?
4. Is your antivirus program configured to check every file on your computer (CD-ROMS, floppy disks, e-mail, and the Web)?
5. Do you have heuristic tests enabled?

19

Choosing and Using
Security Professionals

Although there is much you can do to protect yourself, sometimes it's best to hire security professionals. You probably would save time and money by getting a locksmith to rekey the locks in your home, for example, rather than trying to rekey them yourself. When looking for security *services*, most people pick a company randomly from a local telephone book. But beware: Too many security-related businesses are run by dishonest, self-proclaimed "experts" who have had little security training or experience. At best, relying on an incompetent or dishonest security professional can cost you money unnecessarily. At worst, it can place you and your family in danger. This chapter explains what various security professionals can do for you, how to find good ones, and how to get the most out of them.

In addition to locksmiths, the security professionals you might use include security consultants, self-defense instructors, and installers of home alarms, home automation systems, and car alarms. Before hiring any service provider, make sure that the company or individual has been doing business in your area for a few years and doesn't have a trail of unhappy customers. You often can get this information by calling your Better Business Bureau, state attorney general's office or consumer protection agency, or local Chamber of Commerce. If the company is a corporation, you can find out when it began doing business by calling your state attorney general's corporation department.

Before hiring a security professional, you'll also need to know if the person is competent. The best way to do this is to consider his or her licenses, certifications, memberships, training, experience, and status among professional peers. The questions you'll need to ask vary among types of security professionals.

Locksmiths

Locksmiths primarily sell and service locking devices, safes, and other security hardware. Some also handle home and car alarms. If you just need someone to unlock your car door, you probably don't need to worry about which locksmith you choose. Virtually all locksmiths know how to open most cars. To avoid being charged too much when you are locked out of your car, don't call only the first locksmith listed in your telephone book. Call two or three, and ask each one what the charge will be. Stay away from locksmiths who won't give you a flat fee. The charge should depend on how far away your car is from the locksmith's shop and on your car's make, model, and year—not on how long it takes to unlock your door. Locksmiths who have different skill levels will take different amounts of time to unlock the same vehicle, but you shouldn't have to pay more to a lock-smith who isn't very skilful. (Expect to pay more, however, if new car keys must be made for you.)

Be sure to check out carefully any locksmith who will work at your home. A small percentage of them are former burglars. In most places, a locksmith doesn't need to be licensed. Even where "licensing" is required, an applicant may not have to submit to a criminal background check or meet any competency or training standards. The applicants usually need only sign a business registra-tion form and pay a small fee. New York City and California are among the few places that require applicants for a locksmith license to submit to a background check and meet certain training and competency standards.

Where licensing laws are lax or nonexistent, you'll need to consider a lock-smith's certifications. The most respected locksmith certification program is offered by the Associated Locksmiths of America (ALOA), the oldest and largest nonprofit locksmith trade association. ALOA offers three certification levels to members and nonmembers: Registered Locksmith (RL), Certified Professional Locksmith (CPL), and Certified Master Locksmith (CML). If a locksmith's Yellow Pages advertisement includes ALOA's logo and the initials RL, CPL, or CML, the locksmith is showing certification by the ALOA. Some locksmiths simply advertise themselves as "Master Locksmiths." Don't pay attention to the term because it has no meaning.

Don't be surprised if you can't find an ALOA-certified locksmith near you. Most locksmiths—including many highly respected ones—haven't been tested by ALOA. Some disagree with the organization's legislative activities; others think the testing fees are too high or don't believe that testing is necessary. It's impor-tant to realize that although ALOA certification is an indication of competency, lack of such certification doesn't necessarily indicate lack of competence.

Manufacturers of high-security locks and cylinders also offer certification pro-grams for locksmiths. Although these certifications are based on proficiency with a specific product line, they also indicate overall competency because a locksmith has to have a good foundation of knowledge to understand high-security locks.

Next to certification, the best indicator of competency for a locksmith is mem-bership in certain associations that provide continuing education to their members. You may have seen the various logos in your Yellow Pages under

"Locks & Locksmiths." The best-known national associations for locksmiths are ALOA, the Door and Hardware Institute (DHI), and the National Locksmith Association and Trust (NLA). ALOA and DHI have the toughest membership criteria; membership in the NLA means the locksmith subscribes to the National Locksmith Magazine. The publication helps to keep locksmiths abreast of fast-changing technical and product information.

In addition to national associations (which include members from both the United States and Canada), there are many state and regional locksmith associations—British Columbia Locksmiths Association, Central Pennsylvania Locksmith Association, Locksmith Association of Connecticut, Maryland Locksmiths Association, Master Locksmiths of Quebec, Missouri-Kansas Locksmith Association, Oregon Association of Professional Locksmiths, and many others. The groups vary greatly in their membership criteria and educational benefits. Still, a locksmith who is a member of any of the major locksmith associations is a better risk than someone who doesn't belong to any of them. Generally, if a locksmith belongs to one of the groups, the membership will be stated in the Yellow Pages advertisement.

A locksmith who specializes in safes also may belong to the Safe and Vault Technicians' Association (SAVTA) or the National Safeman's Organization. SAVTA is the more prestigious of the two and is harder to join.

Don't be surprised if you find that few locksmiths in your area are members of any association. Most locksmiths have never joined an association. As with certifications, lack of membership doesn't indicate incompetence, but lack of proper training and experience does.

Locksmiths generally learn their trade in apprenticeship training, correspondence courses, or classroom training. A well-staffed, accredited school provides the broadest base of knowledge. Apprenticeship training can be great, if done under a competent locksmith. Correspondence courses are hard to judge: A person can pass the tests (mainly written exams) without learning much. A locksmith who graduated from an accredited school is a better risk than one who didn't.

Next in importance to how locksmiths learned the trade is how long they've been practicing it. Look for a locksmith who has been working in your area for at least five years.

Home Alarm Systems Installers

Alarm installers generally handle a wide range of electronic security devices, including home alarms, home automation systems, and access-control systems. Some also install car alarms. As with locksmiths, alarm installers should be checked out carefully before they work in your home. The person who installs your electronic security system can easily disable it at any time (or tell someone else how to disable it). Also, a lot can go wrong with an improperly installed system. You'll save yourself and your neighbors a lot of sleepless nights by choosing a competent person to install your system.

Many cities and states license alarm installation companies—that is, the company itself or the owner is often required to be licensed, but the person actually doing the work may not be licensed. Call your state and local licensing agencies to learn about the requirements in your area.

Regardless of the local licensing requirements, both the company and the installer should have been installing security systems for at least five years. Neither should have a bad record with your local or regional Better Business Bureau.

You can be reasonably sure that you are hiring a competent person if your alarm installer has been certified by the National Alarm Association of America (NAAA) or the National Burglar and Fire Alarm Association (NBFAA). Both organizations offer highly respected certification programs for members and non-members. In addition to these two national associations, there are many state and regional groups for alarm installers. As with locksmiths, however, lack of certifications or memberships doesn't necessarily indicate incompetence.

Alarm systems installers usually learn their trade through apprenticeship or classroom training. In most places that license alarm installers, you don't need to worry about where your installer learned the trade. If you want the best alarm installer around, look for one who has been certified by a national trade association *and* is a licensed electrician.

Physical Self-Defense Instructors

Any physical self-defense instructor should have a black belt (or equivalent training) in one or more martial arts and at least three years' experience as an instructor. Certifications and memberships are less important considerations when choosing a self-defense instructor than when choosing other security professionals.

Don't be impressed by how many trophies and plaques an instructor has. Those awards simply mean that he or she has won some martial arts competitions. You're looking for streetwise self-defense, and a martial arts competition no more resembles a street encounter than does a boxing or wrestling match. On the street, you have no padded protective gear, no referees, no timed rounds, and no rules. You want an instructor who knows how to fight and how to teach fighting—not necessarily one who knows how to compete.

Look for someone who emphasizes self-defense techniques that a smaller person can use to defend against a larger person. These include kicks below the waist, open-handed strikes to vulnerable areas of the body, and the use of weapons. Be suspicious of instructors who teach "Bruce Lee style" high kicks. They look great in the movies, but high kicks seldom work well in a street fight. Low kicks provide better balance and more speed and power. (If the instructor is teaching high kicks mainly as an exercise in balance and flexibility, that's fine. But if you are told that such kicks are good for self-defense, find another instructor.) Stay away from instructors who emphasize hardening of the hands and feet, breaking boards with your limbs, or brute-strength fighting techniques. These feats are impractical for most people and take many years to learn.

The best way to judge a physical self-defense instructor is to observe one during a class. All good instructors will be happy to let you watch at no charge. Pay attention to how the instructor interacts with the students. Ask yourself: Do they respect the instructor? Do they understand the instructions? Is the class run in an orderly fashion? Does the instructor know how to fight?

Security Consultants

Although this book contains everything most people need to know about home security, there's much more to the subject than can be included in one book. If you live in an expensive house, keep high-priced valuables in your home, have a high-profile job, or know that people want to harm you or your family, you should consult with a qualified security consultant. This professional can help you to assess your security risks accurately and can show you the most cost-effective ways to meet those risks (Figure 19.1).

Many locksmiths and home alarm installers who call themselves "security consultants" are basically salespersons. Because they mainly want to sell you something, they're likely to suggest products and services that you don't need. In some cities, police officers give free security advice. However, most police officers aren't specialists in using security hardware or electronic alarm systems. The advice they give is very general: "Always use deadbolts;" "Keep your doors locked;" "Keep your windows closed."

For unbiased, accurate, and in-depth advice, you'll need an independent security consultant, someone who has a broad range of security-related training and experience and can show you ways to get the best protection at the lowest cost. Independent security consultants know where to find a wide variety of security products at discount prices. Often they can save you more money than they charge. But, again, beware: Not all independent security consultants are useful.

Figure 19.1 Security consultants can help you to get the most security for the least amount of money.

In most places, anyone can legally call himself or herself a security consultant. No special training or licensing is required. A competent security consultant will have had extensive training and experience in two or more of the following areas: criminology, law enforcement, locksmithing, alarm system installation, safe servicing, martial arts, or firefighting. To find out what training or experience a security consultant has, *ask*. He or she should be happy to provide the information to you in writing (usually in the form of brochures and flyers).

Try to get references before you hire a security consultant. This may be hard because of the confidential nature of the profession. It's not in the best interest of most clients to act as reference sources.

Another indication that a security consultant is qualified is whether he or she has been certified by a security trade association. The most prominent associations are the American Society for Industrial Security (ASIS), the International Association of Home Safety and Security Professionals (IAHSSP), and the International Association of Professional Security Consultants (IAPSC).

ASIS is the oldest and largest trade association for security consultants. It offers the Certified Protection Professional (CPP) certification to members and nonmembers who meet certain guidelines and pass a rigorous test. IAHSSP is the only association that specializes in training security professionals who serve home owners and apartment dwellers. It offers the Residential Protection Specialist (RPS) registration to security professionals who pass a rigorous test. The IAPSC offers the Certified Protection Officer (CPO) status to members and nonmembers who pass a test.

A friend asked me, "If I didn't know you, how would I find the best security consultant around?" This question made me seriously consider how a layperson could distinguish among a group of competent security consultants. A comparison of licenses, certifications, memberships, training, and experience probably wouldn't help much because several of the consultants might appear to be equally qualified.

To find top-notch security consultants, you have to consider their professional reputations among their peers. The most respected consultants are those who hold or have held offices in trade associations, teach security classes at accredited schools and trade association seminars, write books for the trade, or have technical articles published in the major trade journals. Unless you live in a major city, it's unlikely that more than one security consultant in your area will meet one of these criteria. The person who has those credentials is probably the best around.

Preventing and Surviving Home Fires

Vying with Canada during the past two decades, the United States continues to have one of the worst fire death records among industrialized countries. Most fire deaths in North America occur in homes and could have been avoided if the victims had taken simple precautions.

Many people in the United States and Canada don't take fire safety seriously. During school fire drills, for instance, teachers and students stand outside talking and giggling. We tend to feel sympathy for a person who experiences a home fire. In Great Britain and other countries, fire victims are penalized for their carelessness. Perhaps the contrast in attitudes has something to do with the difference in fire death rates.

This chapter looks at how home fires occur, how you can avoid them, and how you and your family can survive one. I'll tell you about some important fire safety devices and show you how they can best be used.

Causes and Cures

According to the U.S. Fire Administration, most home fires can be traced to smoking, cooking, heating equipment, and electrical appliances. More civilians die in fires related to in-house smoking than any other type of fire. Over 90 percent of fire deaths each year are the result of someone falling asleep or passing out while holding a lighted cigarette or while a lighted cigarette was burning out on a nearby furniture surface or in a waste basket. Mattresses, stuffed chairs, and couches often trap burning ashes for long periods of time while releasing poisonous gases. Many people are killed by the gases rather than by flames.

The best way to avoid smoking-related fires is not to smoke in your home. If you're a smoker or allow other people to smoke in your home, be sure that sturdy, deep ashtrays are in every room in which people smoke. Always douse butts with water before dumping them in the trash, and check under and behind cushions for smoldering butts before leaving home or going to bed. Never smoke when you're drowsy or while you're in bed.

The kitchen, where people work with fire most frequently, is the leading room of origin for home fires. Here are some simple things you can do to virtually eliminate the risk of ever having a major kitchen fire:

1. Keep your stove burners, oven, and broiler clean and free of grease.
2. When you're cooking on the stovetop, never leave it unattended.
3. Turn handles of pots and pans away from the edges of the stove while cooking.
4. When you cook, wear short sleeves or keep your sleeves rolled up (to avoid dragging them near the flames).
5. Make sure that no towels, paper, food wrappings or containers, or other flammable items are close to the stove.
6. Don't use towels as pot holders (towels ignite too easily).
7. Never store flammable liquids in the kitchen.

You still may have a small grease fire occasionally while cooking. (Be especially careful when frying foods.) Be prepared to respond immediately to such a fire. Respond to a small grease fire on the stove by turning off all burners of the stove and quickly covering the burning pan with a large metal lid. If no metal lid is at hand, pour a large quantity of flour onto the burning area to smother the flames while you get a cookie sheet that can be placed of the top of the pan to seal off oxygen—or use your fire extinguisher to put the fire out. Don't pick up the pan to carry it to the sink. You may burn yourself, spill burning grease, or drop the pan and start a fire on the floor.

Although more fires start in kitchens than in any other rooms, cooking isn't the main culprit. The number one cause of home fires is heating equipment. Nearly one-fourth of home fires involve space heaters, fireplaces, or wood stoves.

To avoid a heating equipment fire:

- Make sure any heating equipment you buy has been tested and approved by an independent testing laboratory (such as Underwriters Laboratories).

- Be sure to follow manufacturers' instructions when using the equipment.

- Never leave flammable materials near heating equipment.

- If you have a space heater, always keep it at least 36 inches away from anything combustible, including wallpaper, bedding, and clothing.

- At the start of each heating season, make sure that your heating system is in good working order. Check standing heaters for fraying or splitting wires and for overheating. If you notice any problems, have all necessary repairs done by a professional.

During a typical year in the United States, home appliances and wiring problems account for about 100,000 fires and over $760 million in property losses.

Many fires could have been prevented if someone simply had noticed a frayed or cracked electrical cord and had it replaced.

You may think that most of the preceding fire safety suggestions are so obvious that they don't need to be stated. They are "obvious," but everyday fires occur because someone failed to take one of those simple precautions. In addition to following those suggestions, you should have a few safety products, such as smoke detectors and fire extinguishers.

Smoke Detectors

A working smoke detector is the single most important home safety device (Figures 20.1 and 20.2). About 80 percent of all fire deaths occur in homes not equipped with enough working smoke detectors. Most fatal fires happen between midnight and 4:00 A.M., when residents are asleep. Without a working smoke detector, you may not wake up during a fire because the smoke contains poisonous gases that can put you into a deeper sleep.

The vast majority of homes in the United States have at least one smoke detector installed, but most of the detectors don't work because their batteries are dead or missing. Simply having a smoke detector isn't enough. It has to remain in working order to help you and your family stay safe.

There are two basic types of smoke detectors: ionization detectors and photoelectric detectors. They work on different principles, but either type is fine for most homes. Considering that many models sell for less that $10, it's foolish not to have several working smoke detectors in your home.

Smoke detectors should be installed on every level of a home, including the basement. A detector should be placed directly outside each sleeping room. The best location is 6 inches away from air vents and about 6 inches away from walls and corners.

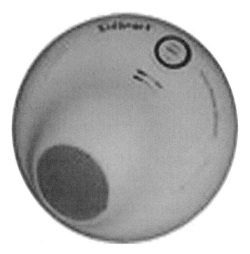

Figure 20.1 A talking smoke detector. (*Courtesy of Signal One.*)

Figure 20.2   A commercial smoke detector.

Test your smoke detectors once a month to make sure that they're in good working order. If they're battery operated, replace the batteries as needed—usually about twice a year. Some models sound an audible alert when the battery is running low. Don't make the mistake of removing your smoke detectors' batteries to use them for operating something else.

The KidSmart vocal smoke alarm

Traditional smoke detectors do not reliably awaken sleeping children. This is so not because the detectors are not loud enough, but rather because our brains respond better to a familiar sound while we are sleeping than to the shrill tone of a conventional alarm. This recently discovered problem has been documented by media stations across the United States.

And the solution—a personally recorded "familiar voice message"—has been studied by respected institutions from around the globe, including the Victoria University Sleep Laboratory of Melbourne, Australia, the world's foremost authority on sleeping and waking behaviors. In those tests, Dr. Dorothy Bruck discovered that 100 percent of all children tested with a "familiar voice" awoke within seconds.

Additionally, tests have either been conducted or are currently ongoing by the following institutions: Consumer Product Safety Commission, U.S. Naval Academy Fire Department, University of Georgia, and others.

Fire Extinguishers

A fire extinguisher can offer good protection if you have the right model and know how to use it. If you use the wrong type, you actually can make a fire spread.

There are several types of fire extinguishers, and each type is designed to extinguish fires from particular sources. The main types of fire extinguishers are

Class A—for wood, paper, plastic, and clothing fires

Class B—for grease, gasoline, petroleum oil, and other flammable liquids fires

Class C—for electrical equipment and wiring fires

For most homes, it's a good idea to buy a class ABC fire extinguisher, because it's useful for a wide range of types of fires.

Buy a fire extinguisher that everyone in your home will find easy to use. It won't be much good if no one is strong enough to lift it. Look for a model that has a pressure-gauge dial. Then you will know at a glance when the pressure is low and the extinguisher needs to be refilled.

When you buy a fire extinguisher, read the instructions carefully. You should be ready to use it correctly and without hesitation at any time. To use most extinguishers, you should stand at least 8 feet away from the fire, remove a pin from the extinguisher, aim the nozzle at the base of the fire, and squeeze the trigger while sweeping the nozzle back and forth at the base of the fire until you're sure that the fire is out. An easy way to remember how to use the fire extinguisher is to remember the acronym *PASS*, which stands for *p*ull, *a*im, *s*queeze, and *s*weep.

Don't ever conclude that because you own a fire extinguisher, you have no need for your local firefighters if a fire breaks out in your home. You aren't a trained firefighter, and the visible flames are only one lethal element of a fire. Unless it's a small fire that you quickly put out, call your fire department. You should use your extinguisher only to put out a small fire and only if the fire is between you and your only means of escape. Many small fires spread quickly and become uncontrollable and life threatening. Before trying to put out any fire, make sure that you have a way to escape, and use it immediately if the extinguisher doesn't put out all the fire.

Escape Ladders

If you live in a multiple-story home, plan a way to escape safely from windows located above the ground floor. You might install rope-ladder hooks outside each upper-floor bedroom and keep a rope ladder in each of the bedroom closets. Another option is to use a fixed ladder, such as the Redi-Exit (Figure 20.3).

The Redi-Exit is a unique ladder that is disguised as a downspout when not being used. Its shape discourages people from trying to use it to gain entry into a home. From an upper-floor window, you can open up the Redi-Exit by striking down on a release knob. The unit can be installed on a new or existing home.

Fire Sprinkler Systems

Studies by the U.S. Fire Administration indicate that the installation of quick-response fire sprinkler systems in homes could save thousands of lives, prevent

Figure 20.3 A Redi-Exit ladder in a closed and an open position. (*Courtesy of REDDCO, Inc.*)

a large portion of fire-related injuries, and eliminate hundreds of millions of dollars in property losses each year. Sprinklers are the most reliable and effective fire protection devices known because they operate immediately and don't rely on the presence or actions of people in the building (Figure 20.4). Residential sprinklers have been used by businesses for over a century, but most homeowners haven't considered installing them because they are misinformed about sprinklers and misunderstand their use.

One misconception about residential sprinklers is that all of them will be activated at once, dousing the entire house. In reality, only the sprinkler directly over the fire will go off because each sprinkler head is designed to react individually to the temperature in that particular room. A fire in your kitchen, for example, won't activate a sprinkler head in your bedroom.

Figure 20.4   A commercial sprinkler head.

Another misconception is that fire sprinklers are prohibitively expensive. A home sprinkler system can cost less than 1 percent of the cost of a new home—about $1.50 per square foot. The additional cost may be minimal when spread over the life of a mortgage. You may find that a home sprinkler system virtually pays for itself in homeowner's insurance savings. Some insurers give up to 15 percent premium discounts for homes with sprinkler systems.

If you can't see your way clear to installing a full home sprinkler system, consider one that protects one of your most vulnerable areas—your kitchen stove. The Guardian is the first automatic range-top fire extinguisher available for home use. It was developed for U.S. military use after a 1984 study identified cooking-grease fires as the number one cause of fire damage and injuries in military-base housing. The patented system uses specially calibrated heat detectors to trigger the release of a fire-extinguishing chemical.

When the chemical is released, the system automatically shuts off the stove. In laboratory tests, The Guardian has been found to detect and extinguish stove-top fires within seconds—but not to activate under normal cooking conditions. You can have it installed so that it also activates an alarm inside your home.

The Guardian is UL listed and combines a fire-detection assembly and a chemical distribution system into a single automatic unit. The fire-detection system can be installed neatly under any standard range-top hood. Cables connect it to the extinguisher assembly, which is housed in the cabinetry above the stove top. A pressurized container stores a fire-extinguishing liquid that is distributed through stainless steel piping to the underhood nozzles.

Here's how The Guardian responds when a stove-top fire starts:

1. Extreme heat from the stove-top fire causes any of four fusible links in the underhood detection assembly to separate, releasing tension on a cable.

2. When the cable tension is released, a tension spring automatically opens the extinguisher valve, discharging the liquid extinguishing mixture through the piping.

3. The mixture flows through two nozzles positioned directly above the stove-top burners, and a controlled discharge smothers the fire and guards against another fire starting.

4. While the extinguishing mixture is being released, a microswitch activates a switch that shuts off the gas or electric fuel source.

Surviving a Home Fire

To ensure that you and your family will be able to get out alive during a fire, you need to plan ahead. All members of the household, including small children, should help to develop an escape plan and regularly practice using it. It isn't enough to just say what you plan to do in case of a fire. You may have only seconds to get out, and the smoke may be so thick and black that you won't be able to see where you're going. Only through practice will you be able to react quickly and do almost routinely what you need to do to survive.

Make sure that all potential escape routes are readily accessible. Check that your windows aren't painted shut. Remove furniture that is blocking a door. Adjust locks that are too high for children to reach. And so on. Take care of any obstacles right away.

Establish a meeting place outside and not too close to your home (a spot near a designated tree or on a neighbor's porch, for instance). Agree that all members of the household will go there and wait together for the fire department. Everyone should know how to call for help—either at a neighbor's home or by using a fire box (Figure 20.5).

No one should go back into a burning home, even if someone is unaccounted for. If you go back into your home, you're endangering not only yourself but also anyone who's inside. Fire grows quickly, and it rushes to wherever there is oxygen. As you open windows or doors in a burning home, you're making the smoke and flames spread faster. It's better to stay outside and wait for the firefighters. They'll arrive quickly and will have the equipment and skills to rescue anyone left inside.

Here are some key actions that everyone should remember. If you encounter smoke on your way out, try to use an alternative exit. If you must escape through smoke, crawl along the floor, under the smoke, where the air is cooler and cleaner.

Figure 20.5   A fire pull box.

If your clothing catches on fire, stop, drop to the ground, and roll to extinguish the flames.

If you're in a bedroom and you hear a smoke detector but don't see smoke, leave quickly through a bedroom window if possible. If the room is too high off the ground or you can't get out of the window safely, feel the door from the bottom up to find out if it's warm. Don't touch the door knob; it may be hot. Don't open the door if it feels warm. If the door is cool, place your shoulder against it and open it slowly. If you don't see flames and an exit is near, quickly crawl to safety. Once you are out of the building, call the fire department immediately. Don't go back into the building for any reason.

If the bedroom door is hot and you can't safely climb out of a window, stuff rags or rolled up clothes under and around the door and in every gap or opening that may allow smoke to enter the room. If you can't climb out of a window safely, hang a rag or piece of clothing out of it. This will let firefighters know where you are.

What to Do after a Fire

If you have a home fire, take these actions as quickly as you can afterward:

1. Immediately call your insurance company or the insurer's agent, and then call your mortgage company.

2. Don't let anyone into your home without first seeing identification. Criminals may try to take advantage of your vulnerable situation.

3. Make sure that all your utilities are turned off. If you're in a cold climate and you expect your house will be empty for a long time, drain the water lines.

4. If possible, board up all holes in the roof, doors, windows, and other entry points.

5. Protect all undamaged property to avoid further damage.

6. Don't clean up until after your insurance company has inspected the damage.

7. Make a list of all your damaged property. If possible, include the model numbers, serial numbers, dates and places of purchase, and purchase prices. The more details you have about your property, the better off you'll be when dealing with your insurance company.

8. If your home is too damaged to live in, and you need temporary shelter, call your insurance company, the local Red Cross, or the Salvation Army for help. Other possible sources of help include churches and synagogues, and civic groups such as the Lions Clubs International and the Rotarians.

9. Keep all receipts for additional living expenses and loss-of-use claims.

Be wary of uninvited insurance adjusters who may contact you after hearing a report of the fire. If you have a complete inventory of your property and it's readily accessible, an insurance adjuster probably can't do any more for you than you can do for yourself.

Chapter 2

Terrorism

We remain a nation at war. I wish I could report,
you know, a different sentence to you. But my job
as the President of the United States is to keep the
American people fully informed of the world in
which we live. In recent months, I've spoken
extensively about our strategy for victory in Iraq.
Today, I'm going to give you an update on the
progress that we're making in the broader war on
terror: The actions of our global coalition to break
up terrorist networks across the world, plots we've
disrupted that have saved American lives, and
how the rise of freedom is leading millions to reject
the dark ideology of the terrorists—and laying the
foundation of peace for generations to come.
    PRESIDENT GEORGE W. BUSH, FEBRUARY 9, 2006

Terrorism is the use or threat of violence to create fear and alarm for political
or religious purposes. Terrorists murder and kidnap people, set off bombs, hijack
airplanes, set fires, and commit other serious crimes. Despite the huge war
machine of the United States, we will never have the firepower to rid the world
of terrorism. Terrorists hide in the shadows and lurk in the alleys of the world,
and with countless miles of places to hide in, it is impossible to stop all terrorists.

The Camden County, Georgia, Sheriff's Office Web site offers the following
information:

Terrorism doesn't mean you have to change your life—it just means that you
need to be prepared. Whether a disaster is natural or human made, it's important to be prepared. Meet with your family and discuss why you need to be prepared for a disaster, and work together to prepare a family disaster plan.

Discuss the types of hazards that could affect your family. Determine escape
routes from your home and places to meet. Pick places for your family to meet
outside your home in case of a sudden emergency, such as a fire, or outside your
neighborhood if you can't return home.

Have an out-of-state friend or relative as the family contact so that all your family members have a single point of contact. Family members need to call this family contact to let them know where they are in the event you cannot be together.

Make a plan now for what to do with your pets if you need to evacuate.

Post emergency telephone numbers by your phones or in your wallet or purse, and make sure that your children know how and when to call 911.

Stock nonperishable emergency supplies and a disaster supply kit. [*Source*: American Red Cross.]

There are six basics you should stock for your home: water, food, first-aid supplies, clothing and bedding, tools and emergency supplies, and special items. Keep the items in an easy-to carry container, such as a covered trash container, backpack, or duffle bag.

What Is Suspicious Activity?

Residents may observe a variety of actions, statements, associations, or timing or patterns of activity that create suspicions of illegal conduct. No one has a better perspective about what defines "normal" in Georgia communities than the people who live there. Law Enforcement has long relied upon the common-sense perceptions of Georgia citizens who notice something or someone that appears suspicious or out of place.

Who Should I Call to Make a Report?

Call our local police or sheriff or the nearest Georgia State Patrol Post. Tell the operator that you want to make a suspicious activity report. Ask the operator to alert Georgia Homeland Security. Your local law enforcement agency will contact the Georgia Information Sharing and Analysis Center called GISAC. Agents from the Georgia Bureau of Investigation or the FBI will be assigned to carefully check out your information.

Should I Give Police My Name and Contact Number?

Yes! If you want your report to be taken seriously, you should be willing to give your name and contact information to investigators. Someone from Georgia Homeland Security will want to talk to you personally in order to understand the full details of your information and take appropriate action in a timely manner.

Will My Identity Be Protected?

Yes! Reports to Georgia Homeland Security are considered an important part of America's ongoing investigation into the war on terrorism. Investigators will need to know your name and contact numbers in order to do their job, but the State of Georgia will make every effort to keep your identity confidential.

Do I Have to Talk to the News Media?

No! No one who makes a report to Georgia Homeland Security is required to speak with the news media. Georgia Homeland Security will not release your name to reporters. The decision to remain anonymous to the public or to speak with the news media is left completely up to you.

How Should I Focus My Attention?

Everyone should be especially mindful of suspicious activity around what Homeland Security calls "critical infrastructure." These sites are places or facilities where damage or destruction could cause an interruption of service or result in serious injury or death.

What Should I Watch For?

Georgians should immediately report people who photograph, videotape, sketch, or seek blueprints for dams, drinking water supplies, and water treatment facilities; major highway intersections, bridges, and tunnels; ports, transportation hubs, airports and shipping facilities; electric plants and substations and nuclear facilities and transmission towers; pipelines and tank farms; military installations, law enforcement agencies, and defense contract sites; hospitals and health research facilities; Internet, phone, cable, and communications facilities and towers; and capitol, court, and government buildings. Suspicious activity around historic structures and national landmarks also should be reported.

Is My Awareness Really That Important?

Intelligence agents at the Georgia Information Sharing and Analysis Center (GISAC) have investigated an average of one Homeland Security tip every day for nearly a year since the 9/11 attacks. Homeland Security Coordinator Robert Hightower says that he is proud of the many Georgians who have helped to keep the state safe from attacks by reporting suspicious activity.

Homeland Security Presidential Directive

Purpose

The nation requires a Homeland Security Advisory System to provide a comprehensive and effective means to disseminate information regarding the risk of terrorist acts to Federal, State, and local authorities and to the American people. Such a system would provide warnings in the form of a set of graduated

"Threat Conditions" that would increase as the risk of the threat increases. At each Threat Condition, Federal departments and agencies would implement a corresponding set of "Protective Measures" to further reduce vulnerability or increase response capability during a period of heightened alert.

This system is intended to create a common vocabulary, context, and structure for an ongoing national discussion about the nature of the threats that confront the homeland and the appropriate measures that should be taken in response. It seeks to inform and facilitate decisions appropriate to different levels of government and to private citizens at home and at work.

Homeland Security

Since September 11, 2001, President Bush has restructured and reformed the Federal government to focus resources on counterterrorism and ensure the security of our homeland.

Homeland Security Advisory System

The Homeland Security Advisory System shall be binding on the executive branch and suggested, although voluntary, to other levels of government and the private sector. There are five Threat Conditions, each identified by a description and corresponding color. From lowest to highest, the levels and colors are:

Low = Green;

Guarded = Blue;

Elevated = Yellow;

High = Orange;

Severe = Red.

The higher the Threat Condition, the greater the risk of a terrorist attack. Risk includes both the probability of an attack occurring and its potential gravity. Threat Conditions shall be assigned by the Attorney General in consultation with the Assistant to the President for Homeland Security. Except in exigent circumstances, the Attorney General shall seek the views of the appropriate Homeland Security Principals or their subordinates, and other parties as appropriate, on the Threat Condition to be assigned.

Threat Conditions may be assigned for the entire Nation, or they may be set for a particular geographic area or industrial sector. Assigned Threat Conditions shall be reviewed at regular intervals to determine whether adjustments are warranted.

For facilities, personnel, and operations inside the territorial United States, all Federal departments, agencies, and offices other than military facilities shall conform their existing threat advisory systems to this system and henceforth administer their systems consistent with the determination of the Attorney General with regard to the Threat Condition in effect.

The assignment of a Threat Condition shall prompt the implementation of an appropriate set of Protective Measures. Protective Measures are the specific steps an organization shall take to reduce its vulnerability or increase its ability to respond during a period of heightened alert. The authority to craft and implement Protective Measures rests with the Federal departments and agencies. It is recognized that departments and agencies may have several preplanned sets of responses to a particular Threat Condition to facilitate a rapid, appropriate, and tailored response. Department and agency heads are responsible for developing their own Protective Measures and other antiterrorism or self-protection and continuity plans, and resourcing, rehearsing, documenting, and maintaining these plans. Likewise, they retain the authority to respond, as necessary, to risks, threats, incidents, or events at facilities within the specific jurisdiction of their department or agency, and, as authorized by law, to direct agencies and industries to implement their own Protective Measures. They shall continue to be responsible for taking all appropriate proactive steps to reduce the vulnerability of their personnel and facilities to terrorist attack. Federal department and agency heads shall submit an annual written report to the President, through the Assistant to the President for Homeland Security, describing the steps they have taken to develop and implement appropriate Protective Measures for each Threat Condition. Governors, mayors, and the leaders of other organizations are encouraged to conduct a similar review of their organizations' Protective Measures.

The decision whether to publicly announce Threat Conditions shall be made on a case-by-case basis by the Attorney General in consultation with the Assistant to the President for Homeland Security. Every effort shall be made to share as much information regarding the threat as possible, consistent with the safety of the nation. The Attorney General shall ensure, consistent with the safety of the nation, that State and local government officials and law enforcement authorities are provided the most relevant and timely information. The Attorney General shall be responsible for identifying any other information developed in the threat assessment process that would be useful to State and local officials and others and conveying it to them as permitted consistent with the constraints of classification. The Attorney General shall establish a process and a system for conveying relevant information to Federal, State, and local government officials, law enforcement authorities, and the private sector expeditiously.

The Director of Central Intelligence and the Attorney General shall ensure that a continuous and timely flow of integrated threat assessments and reports is provided to the President, the Vice President, Assistant to the President and Chief of Staff, the Assistant to the President for Homeland Security, and the Assistant to the President for National Security Affairs. Whenever possible and practicable, these integrated threat assessments and reports shall be reviewed and commented upon by the wider interagency community.

A decision on which Threat Condition to assign shall integrate a variety of considerations. This integration will rely on qualitative assessment, not quantitative calculation. Higher Threat Conditions indicate greater risk of a terrorist act, with risk including both probability and gravity. Despite best efforts, there can be no guarantee that, at any given Threat Condition, a terrorist attack will not occur. An initial and important factor is the quality of the threat information itself. The evaluation of this threat information shall include, but not be limited to, the following factors:

1. To what degree is the threat information credible?
2. To what degree is the threat information corroborated?
3. To what degree is the threat specific and/or imminent?
4. How grave are the potential consequences of the threat?

Threat Conditions and Associated Protective Measures

The world has changed since September 11, 2001. We remain a nation at risk to terrorist attacks and will remain at risk for the foreseeable future. At all Threat Conditions, we must remain vigilant, prepared, and ready to deter terrorist attacks. The following Threat Conditions each represent an increasing risk of terrorist attacks. Beneath each Threat Condition are some suggested Protective Measures, recognizing that the heads of Federal departments and agencies are responsible for developing and implementing appropriate agency-specific Protective Measures:

1. *Low condition (green).* This condition is declared when there is a low risk of terrorist attacks. Federal departments and agencies should consider the following general measures in addition to the agency-specific Protective Measures they develop and implement:

   Refining and exercising as appropriate preplanned Protective Measures;
   Ensuring personnel receive proper training on the Homeland Security Advisory System and specific preplanned department or agency Protective Measures; and
   Institutionalizing a process to assure that all facilities and regulated sectors are regularly assessed for vulnerabilities to terrorist attacks, and all reasonable measures are taken to mitigate these vulnerabilities.

2. *Guarded condition (blue).* This condition is declared when there is a general risk of terrorist attacks. In addition to the Protective Measures taken in the previous Threat Condition, Federal departments and agencies should consider the following general measures in addition to the agency-specific Protective Measures that they will develop and implement:

   Checking communications with designated emergency response or command locations;
   Reviewing and updating emergency response procedures; and
   Providing the public with any information that would strengthen its ability to act appropriately.

3. *Elevated condition (yellow).* An Elevated Condition is declared when there is a significant risk of terrorist attacks. In addition to the Protective Measures taken in the previous Threat Conditions, Federal departments and agencies

should consider the following general measures in addition to the Protective Measures that they will develop and implement:

Increasing surveillance of critical locations;
Coordinating emergency plans as appropriate with nearby jurisdictions;
Assessing whether the precise characteristics of the threat require the further refinement of preplanned Protective Measures; and
Implementing, as appropriate, contingency and emergency response plans.

4. *High condition (orange).* A High Condition is declared when there is a high risk of terrorist attacks. In addition to the Protective Measures taken in the previous Threat Conditions, Federal departments and agencies should consider the following general measures in addition to the agency-specific Protective Measures that they will develop and implement:

Coordinating necessary security efforts with Federal, State, and local law enforcement agencies or any National Guard or other appropriate armed forces organizations;
Taking additional precautions at public events and possibly considering alternative venues or even cancellation;
Preparing to execute contingency procedures, such as moving to an alternate site or dispersing their workforce; and
Restricting threatened facility access to essential personnel only.

5. *Severe condition (red).* A Severe Condition reflects a severe risk of terrorist attacks. Under most circumstances, the Protective Measures for a Severe Condition are not intended to be sustained for substantial periods of time. In addition to the Protective Measures in the previous Threat Conditions, Federal departments and agencies also should consider the following general measures in addition to the agency-specific Protective Measures that they will develop and implement:

Increasing or redirecting personnel to address critical emergency needs;
Assigning emergency response personnel and pre positioning and mobilizing specially trained teams or resources;
Monitoring, redirecting, or constraining transportation systems; and
closing public and government facilities.

Comment and Review Periods

The Attorney General, in consultation and coordination with the Assistant to the President for Homeland Security, shall, for 45 days from the date of this directive, seek the views of government officials at all levels and of public interest groups and the private sector on the proposed Homeland Security Advisory System.

One hundred thirty-five days from the date of this directive the Attorney General, after consultation and coordination with the Assistant to the President for Homeland Security, and having considered the views received during the comment period, shall recommend to the President in writing proposed refinements to the Homeland Security Advisory System.

22

Creating a Web Site

Many locksmithing shops and security companies have a Web site. It can help you to educate potential customers, promote all your products and services, and make online sales. You can even include a map to make it easier for customers to find you. This chapter tells you everything you need to know to create your own Web site or work with a professional Web site designer.

When someone gives you his or her Web address, it generally takes you to that person's Web site's home page, which should introduce you to the information and services that the site offers. The home page is usually the first page of the site. From the home page, you can click on links, which may be graphics or text, that take you to other pages in the site.

Some sites, such as the *National Locksmith*'s, use a *splash page*. This is a Web page that shows up before taking you to the home page. If you type in the domain name TheNationalLocksmith.com, you'll see the splash page. A few seconds later you'll be at the home page. Generally, a splash page isn't necessary. But if you use one, it should automatically load the home page quickly. The longer it takes for someone to navigate your Web site, the more likely they will get frustrated and move on to another site. From the home page, you should be able to click on a graphic or text link to go to any of the main Web pages.

Getting Ideas for Making a Web Site

Before you begin creating your Web site, take a look at other security-related Web sites. They will give you ideas about what to do and what not to do. To search for security Web sites, go to a Web search engine such as google.com. At the search engine, enter "lock and key," "safe and lock," and "lock and safe." Each of those entries will find many locksmithing sites. Some of the sites are professionally done; many look amateurish.

Some home pages have a busy background that makes it hard to read. A plain background is better. You'll also notice that the better Web sites have a good balance between graphics and text. Nice photos and drawings make a Web page inviting to read. You can use a digital camera to take photos that you can use directly on your Web pages. If you want to use print photographs or drawings, you'll need to use a scanner to digitize them first.

If you can't take good photographs, use graphics from product manufacturers. You may need to ask permission first, but most manufacturers won't mind letting you use their images to illustrate your Web site. Some manufacturers, such as Adesco Safe (adesco.com), provide artwork on their Web sites for security dealers to use.

Choosing a Web Site Host

To post your Web site, you'll need to use a Web site host. You can pay for the service or use a free host, such as freeservers.com, hostsltd.com, and free. prohosting.com. For a list of free hosting services, go to google.com and enter "free Web hosting." If you use a free service, you'll need to let the host post advertising banners on your Web pages. You also may be limited in the domain name you can use. You may have to use a subdomain name that identifies the host. Free servers let you use your own domain name, and for a fee, you can avoid the advertising banners.

Choosing a Domain Name

It's important to have an easy-to-remember domain name that's easy to spell. If your company name isn't being used on the Internet, consider buying it from networksolutions.com. At that site you also can check to see if a domain name is available. If you can't or don't want to use your shop name, use one that relates to your location or to the services or products you offer, for example, atlantalocksmith.com or houstonlocksandsafes.com.

In addition to the domain name, you'll need to choose a domain name extension. There are dozens of extensions, such as org, edu, gov, info, net, tv, name, com, and many others. Each extension is used for different types of sites. Gov is used for government sites. Org is used for organizations. Edu is used for educational sites—such as colleges. Net is for sites about the Internet. It's best to use com because that's most often used for business or commercial purposes. Different domain name extensions create different domain names. Thelocksmith. com is different from thelocksmith.net, for instance. If you go to networksolutions. com, you'll be able to search for domain names under many domain name extensions.

Creating Content

If you want people to come to your Web site, you need to create interesting content and update it regularly. Some subjects you may want to cover include securing your home, buying locks, buying a safe, securing your car, and travel safety. Don't make the mistake of having a Web site that has little useful information. Ask yourself what kinds of things your potential customers would want to read about.

Sketching Out Your Web Site

To make a good Web site, you need to plan it before creating Web pages. First, make a sketch beginning with your home page and showing each of the Web pages you will include. The sketch should show how each page connects to the other pages. Every page should be accessible with no more than three clicks. You can do that if you include all the major links on your home page.

Music

It's generally best to not use music on your Web site because it can become annoying quickly, especially if it's played automatically on every Web page. If you use music, do so sparingly, such as only on your splash page or home page. Also, include the option of turning the music off.

Graphics

Don't use a lot of large graphics because they result in it taking a long time to load the Web page. You want your Web pages to load quickly. The longer it takes for them to load, the less likely people will stay on your Web site. Save your photographs in .jpeg or .jpg formats and your drawings in .gif format. If your images are measured in megabytes (MB), they're too large. Use images that are 50 kilobytes (kB) or smaller; they load faster. Use an image editing program, such as Paintshop Pro, to make your images smaller and to choose the format to save your images in.

Using Web Site Creation Software

There are many Web site creation programs. A lot are shareware, which means that you can download them and use them for free (some require you to register them and pay a fee after a free trial period). To find shareware programs, go to google.com or another Web search engine and enter "Web site creation program." Some popular programs include AOL Press, Hotdog, MacroMedia Dreamweaver (free for 30 days), NVU, Paintshop Pro, and Web Express. Paintshop Pro has Web site templates but is most useful for creating and editing images.

One of the most versatile and popular Web site creation programs is Microsoft FrontPage. It includes many templates and wizards and lets you build a Web site that integrates multimedia elements—including sound, video, and discussion pages. Before you buy FrontPage, make sure that your Web host is capable of using it. Some free Web hosts charge a fee to use FrontPage.

You also can find easy-to-use templates on many Web site hosts, such as Freeservers.com. They let you create a Web site just by writing your own text over its text and replacing its graphics with your own. Templates make it easy to create your own professional-looking Web sites.

Helping People Find Your Website

There are thousands of Web sites that relate to security. You need to do certain things to make it easy for people to find your Web site.

Using Search Engines

There are hundreds of search engines on the Internet. People use them to find Web sites. To make your Web site easy to find, submit your Web site information to as many search engines as possible. You can find them by entering a search for "online search engines" and "Internet search engines" at google.com or at another search engine. You also may want to check out Search Engine Watch at searchenginewatch.com to find an exhaustive list of search engines.

Some of the most popular search engines include Altavista (altavista.com), AOL (aolsearch.aol.com), AskJeeves (askjeeves.com), Hotbot (hotbot.com), Looksmart (looksmart.com), Lycos (lycos.com), MSN (search.msn.com), Netscape (search.netscape.com), Open Directory (dmoz.org), and Yahoo! (yahoo.com). On the home page of many search engines you'll find "submit a site," "suggest a URL," "how to suggest a site," and the like. This is where you enter your Web site information for that search engine. There are also Web sites that you can use to submit your information to several search engines at once, such as submitexpress.com/submit.html, website-submission.com, submitasite.com, and submityoururl.info.

If someone enters, say, "locksmiths," "locks," or "safes" in a search engine, thousands of domain addresses will appear. If your Web site appears 1000 entries down the list, for instance, people aren't likely to take the time to click on yours. People generally choose to click on sites that are near the top of the list. This is why you want your site to be as close to the top as possible. There are several things you can do to improve your Web sites' placement in searches, including naming all of your pages and using metatags.

Name Your Pages

Before creating your Web site, create a short, descriptive title for each page. Your site will seem amateurish if you use the default "untitled1.htm," "untitled2.htm," "untitled3.htm," etc. It's better to use titles such as "locks.htm," "safes.htm," "alarms.htm," and the like. Using descriptive titles for each page makes it easier for people to find your Web site through search engines.

Use Meta Tags

If you use Web site creation software, you won't need to know a lot of code—especially if you're using a "what you see is what you get" (WYSIWYG) program. Before submitting your site to a search engines, use metatags at the top of the code portion of each page. When a Web site is uploaded, the codes, including metatags, are invisible. While you're at the code view of your Web site, insert metatags at the top. Many search engines use them to help determine your Web site placement.

Web site creation software makes it easy to switch between viewing the codes and viewing the finished Web page. Metatags go at the top of the code page between <HEAD> and </HEAD>. There are many metatags you can use, but the main ones to include are title, description, and keywords. You want to use words that someone might use to search for your site. Here's an example:

```
<CS>
<HTML>
<HEAD>
<TITLE>Bills Locksmithing shop</TITLE>
<META NAME="keywords" CONTENT="locks, locksmithing, locksmiths, keys, safes, vaults, safe deposit boxes, burglary, crime, security, safety."
<META NAME="description" CONTENT="Locksmithing shop that sells and services locks, keys, safes, alarms, and home, office, commercial, and personal security products."
</HEAD>
</CS>
```

Although they aren't necessary, there are many metatag-generation programs on the Internet. They make it easy to make extensive metatags. You can find some at: http://vancouver-webpages.com/META/mk-metas.html, www.submitcorner.com/Tools/Meta/, and www.netmechanic.com.

Other Ways to Promote Your Site

In addition to getting listed by search engines, you need to find as many ways as possible to promote your Web site. Include your domain address in your advertisements, in your Yellow Pages listing, and on your letterhead and business cards. When you repaint your service vehicles, include your Web site address.

Registered Professional Locksmith Test

This test is based on the International Association of Home Safety and Security Professionals' Registered Professional Locksmith Registration Program. If you earn a passing score, you should be able to pass other locksmithing certification and licensing examinations. To receive a Registered Professional Locksmith certificate, see the information after the test.

1. An otoscope can be helpful for reading disk-tumbler locks by providing light and magnification. *a.* True *b.* False

2. Kwikset locks come with a KW1 keyway. *a.* True *b.* False

3. Many Schlage locks come with an SCL1 keyway. *a.* True *b.* False

4. The purpose of direct (uncoded) codes on locks is to obscure the lock's bitting numbers. *a.* True *b.* False

5. A skeleton key can be used to open warded bit-key locks. *a.* True *b.* False

6. Typically, the lock on a car's driver side will be harder to pick open than other less-often-used locks on the car. *a.* True *b.* False

7. A standard electromagnetic lock includes a rectangular electromagnet and a rectangular wood and glass strike plate. *a.* True *b.* False

8. A blank is a key that fits two or more locks. *a.* True *b.* False

9. One difference between a bit key and a barrel key is that the barrel key has a hollow shank. *a.* True *b.* False

10. Parts of a flat key include the bow, blade, and throat cut. *a.* True *b.* False

11. The Egyptians are credited with inventing the first lock to be based on the locking principle of today's pin-tumbler lock. *a.* True *b.* False

12. Before impressioning a pin-tumbler cylinder, it is usually helpful to lubricate the pin chambers thoroughly. *a.* True *b.* False

13. When you are picking a pin-tumbler cylinder, spraying a little lubrication into the keyway may helpful. *a.* True *b.* False

14. If a customer refuses to pay me after I finish a job at his or her house, I have the legal right to remain inside the house until the person pays me. *a.* True *b.* False

15. A long-reach tool and wedge are used commonly to open locked automobiles. *a.* True *b.* False

16. It's legal for locksmiths to duplicate a U.S. post office box key at the request of the box renter—if the box renter shows a current passport or driver's license. *a.* True *b.* False

17. The Romans are credited with inventing the warded lock. *a.* True *b.* False

18. Five common keyway groove shapes are left angle, right angle, square, V, and round. *a.* True *b.* False

19. To pick open a pin-tumbler cylinder, you usually need a pick and a torque wrench. *a.* True *b.* False

20. Fire-rated exit devices usually have dogging. *a.* True *b.* False

21. Common door lock backsets include
    a. $2^3/_4$ and $2^3/_8$ inches.
    b. $2^3/_8$ and $4^3/_4$ inches.
    c. $1^3/_4$ and $2^1/_4$ inches.
    d. $2^3/_3$ and $3^1/_2$ inches

22. How many sets of pin tumblers are in a typical pin-tumbler house door lock?
    a. 3 or 4
    b. 5 or 6
    c. 11 or 12
    d. 7 or 8

23. Which lock is unpickable?
    a. A Medeco biaxial deadbolt
    b. A Grade 2 Titan
    c. The Club steering wheel lock
    d. None of the above

24. Which are basic parts of a standard key cutting machine?
    a. A pair of vises, a key stop, and a grinding stylus
    b. Two cutter wheels, a pair of vises, and a key shaper
    c. A pair of vises, a key stylus, and a cutter wheel:
    d. A pair of styluses, a cutter wheel, and a key shaper

25. What are two critical dimensions for code cutting cylinder keys?

    a. Spacing and depth
    b. Bow size and blade thickness
    c. Blade width and keyhole radius
    d. Shoulder width and bow size

26. Which manufacturer is best known for its low-cost residential key-in-knob locks?

    a. Kwikset Corporation
    b. Medeco Security Locks
    c. The Key-in-Knob Corporation
    d. ASSA

27. The most popular mechanical lock brands in the United States include

    a. Yale, Master, Corby, and Gardall.
    b. Yale, Kwikset, Master, and TuffLock.
    c. Master, Weiser, Kwikset, and Schlage.
    d. Master, Corby, Gardall, and Tufflock.

28. A mechanical lock that is operated mainly by a pin-tumbler cylinder is commonly called a

    a. disk-tumbler pinned lock.
    b. cylinder-pin lock.
    c. mechanical cylinder-pin lock.
    d. pin-tumbler cylinder lock.

29. A key-in-knob lock whose default position is that both knobs are locked and require that a key be used for unlocking is

    a. a classroom lock.
    b. a function lock.
    c. an institution lock.
    d. a school lock.

30. Four basic types of keys are

    a. barrel, flat, bow, and tumbler.
    b. cylinder, flat, warded, and V-cut.
    c. dimple, angularly bitted, corrugated, and blade.
    d. cylinder, flat, tubular, and barrel.

31. The two most common key stops are

    a. blade and V-cut.
    b. shoulder and tip.
    c. bow and blade.
    d. keyway grooves and bittings.

32. Bit keys most commonly are made of

    a. brass, copper, and silver.
    b. aluminum, iron, and silver.

    c. iron, brass, and aluminum.

    d. copper, silver, and aluminum.

33. Which of the following key combinations provides the most security?

    a. 55555

    b. 33333

    c. 243535

    d. 35353

34. Which of the following key combinations provides the least security?

    a. 243535

    b. 1111

    c. 321231

    d. 22222

35. A blank is basically just

    a. a change key with cuts on one side only.

    b. an uncut or uncombinated key.

    c. any key with no words or numbers on the bow.

    d. a master key with no words or numbers on the bow.

36. You often can determine the number of pin stacks or tumblers in a cylinder by

    a. its key-blade length.

    b. its key-blade thickness.

    c. the key-blank manufacturer's name on the bow.

    d. the material of the key.

37. Spool and mushroom pins

    a. make keys easier to duplicate.

    b. can hinder normal picking attempts.

    c. make a lock easier to pick.

    d. make keys harder to duplicate.

38. As a general rule, General Motors' 10-cut wafer sidebar locks have

    a. a sum total of cut depths that must equal an even number.

    b. up to four of the same depth cut in the 7, 8, 9, and 10 spaces.

    c. a maximum of five number 1 depths in a code combination.

    d. at least one 4-1 or 1-4 adjacent cuts.

39. When drilling open a standard pin-tumbler cylinder, position the drill bit

    a. at the first letter of the cylinder.

    b. at the shear line in alignment with the top and bottom pins.

    c. directly below the bottom pins.

    d. directly above the top pins.

40. When viewed from the exterior side, a door that opens inward and has hinges on the right side is a

    a. left-hand door.

    b. right-hand door.

    c. left-hand reverse-bevel door.

    d. right-hand reverse-bevel door.

41. A utility patent

    a. relates to a product's appearance, is granted for 14 years, and is renewable.

    b. relates to a product's function, is granted for 17 years, and is nonrenewable.

    c. relates to a product's appearance, is granted for 17 years, and is renewable.

    d. relates to a product's function, is granted for 35 years, and is nonrenewable.

42. To earn a UL-437 rating, a sample lock must

    a. pass a performance test.

    b. use a patented key.

    c. use hardened-steel mounting screws and mushroom and spool pins.

    d. pass an attack test using common hand and electric tools such as drills, saw blades, puller mechanisms, and picking tools.

43. Tumblers are

    a. small metal objects that protrude from a lock's cam to operate the bolt.

    b. fixed projections on a lock's case.

    c. small pins, usually made of metal, that move within a lock's case to prevent unauthorized keys from entering the keyhole.

    d. small objects, usually made of metal, that move within a lock cylinder in ways that obstruct a lock's operation until an authorized key or combination moves them into alignment.

44. Electric switch locks

    a. are mechanical locks that have been modified to operate with battery power.

    b. complete and break an electric current when an authorized key is inserted and turned.

    c. are installed in metal doors to give electric shocks to intruders.

    d. are mechanical locks that have been modified to operate with alternating-current (ac) electricity instead of with a key.

45. A popular type of lock used on GM cars is

    a. a Medeco pin tumbler.

    b. an automotive bit key.

    c. a sidebar wafer.

    d. an automotive tubular key.

46. When cutting a lever-tumbler key by hand, the first cut should be the

    a. lever cut.

    b. stop cut.

    c. throat cut.

    d. tip cut.

47. How many possible key changes does a typical disk-tumbler lock have?

    a. 1,500
    b. 125
    c. A trillion
    d. 25

48. Which manufacturer is best known for its interchangeable core locks?

    a. Best Lock
    b. Kwikset Corporation
    c. Ilco Interchangeable Core Corporation
    d. Interchangeable Core Corporation

49. James Sargent is famous for:

    a. inventing the Sargent key-in-knob lock.
    b. inventing the time lock for banks.
    c. inventing the double-acting lever-tumbler lock.
    d. being the first person to pick open a Medeco biaxial cylinder.

50. Which are common parts of a combination padlock?

    a. Shackle, case, bolt
    b. Spacer washer, top pins, cylinder housing
    c. Back cover plate, case, bottom pins
    d. Wheel pack base plate, wheel pack spring, top and bottom pins.

51. General Motors' ignition lock codes generally can be found

    a. on the ignition lock.
    b. on the passenger-side door.
    c. below the Vehicle Identification Number (VIN) on the vehicle's engine.
    d. under the vehicle's brake pedal.

52. Which code series is commonly used on Chrysler door and ignition locks?

    a. EP 1-3000
    b. CHR 1-5000
    c. CRY 1-4000
    d. GM 001-6000

53. How many styles of lock pawls does General Motors use in its various car lines?

    a. One
    b. Five
    c. Over five
    d. Three

54. The double-sided (or 10-cut) Ford key

    a. has five cuts on each side; one side operates the trunk and door, whereas the other side operates the ignition.
    b. has five cuts on each side; either side can operate all locks of a car.
    c. has 10 cuts on each side; one side operates the trunk and door, whereas the other side operates only the ignition.
    d. has 10 cuts on each side.

55. Usually the simplest way to change the combination of a double-bitted cam lock is to
    a. rearrange the positions of two or more tumblers.
    b. remove two tumblers and replace them with new tumblers.
    c. remove the tumbler assembly and replace it with a new one.
    d. connect a new tumbler assembly to the existing one.

56. When shimming a cylinder open
    a. use the key to insert the shim into the keyway.
    b. insert the shim into the keyway without the key.
    c. insert the shim along the left side of the cylinder housing.
    d. insert the shim between the plug and cylinder housing between the top and bottom pins.

57. A lock is any
    a. barrier or closure that restricts entry.
    b. fastening device that allows a person to open and close a door, window, cabinet, drawer, or gate.
    c. device that incorporates a bolt, cam, shackle, or switch to secure an object—such as a door, drawer, or machine—to a closed, locked, on, or off position and that provides a restricted means—such as a key or combination—of releasing the object from that position.
    d. device or object that restricts entry to a given premise.

58. Which wheel in a safe lock is closest to the dial?
    a. Wheel 1
    b. Wheel 2
    c. Wheels
    d. Wheel 0

59. Which is not a type of safe combination wheel?
    a. Hole change
    b. Dial change
    c. Key change
    d. Screw change

60. Which type of cylinder is typically found on an interlocking deadbolt (or jimmy-proof deadlock)?
    a. Mortise cylinder
    b. Key-in-knob cylinder
    c. Rim cylinder
    d. Tubular deadbolt cylinder

Registered Professional Locksmith Answer Sheet

Make a photocopy of this answer sheet to mark your answers on.

1. _____   21. _____   41._____
2. _____   22. _____   42._____

3. ____	23. ____	43. ____
4. ____	24. ____	44. ____
5. ____	25. ____	45. ____
6. ____	26. ____	46. ____
7. ____	27. ____	47. ____
8. ____	28. ____	48. ____
9. ____	29. ____	49. ____
10. ____	30. ____	50. ____
11. ____	31. ____	51. ____
12. ____	32. ____	52. ____
13. ____	33. ____	53. ____
14. ____	34. ____	54. ____
15. ____	35. ____	55. ____
16. ____	36. ____	56. ____
17. ____	37. ____	57. ____
18. ____	38. ____	58. ____
19. ____	39. ____	59. ____
20. ____	40. ____	60. ____

To receive your Registered Professional Locksmith certificate, just submit your answers to this test (a passing score is 70 percent), and enclose a check or money order for US$50 (nonrefundable) payable to IAHSSP. And enclose copies of any two of the following items (do not send original documents because they will not be returned):

- City or state locksmith license
- Driver's license or state-issued identification
- Locksmith suppliers invoice
- Certificate from locksmithing or security school or program
- Yellow Pages listing
- Business card or letterhead from your company
- Locksmithing association membership card or certificate
- Locksmithing or safe technician bond card or certificate
- Letter from your employer or supervisor on company letterhead stating that you work as a locksmith or security professional
- Letter of recommendation from a Registered Professional Locksmith or Registered Security Professional
- Copy of an article you've had published in a locksmith trade journal
- ISBN number and title of locksmith-related book you wrote.

Send everything to IAHSSP, Box 2044, Erie, PA 16512-2044. Please allow 6 to 8 weeks for processing.

Name: _____ Title:

_____

Company name:

_____

_____

Address:

_____

_____

City, State, Zip Code:

_____

Telephone number: _____ E-mail address:

_____

Registered Security Professional Test

This test is based on the International Association of Home Safety and Security Professionals' Registered Security Professional Registration Program. If you earn a passing score, you should be able to pass other security certification and licensing examinations. To receive a Registered Security Professional certificate, see the information after the test.

1. City codes often dictate the height and style of fences. *a.* True *b.* False

2. A hollow-core door is easy to break through. *a.* True *b.* False

3. In general, peepholes should be installed on windowless exterior doors. *a.* True *b.* False

4. A high-security strike box (or box strike) makes a door harder to kick in than a standard strike plate. *a.* True *b.* False

5. A skeleton key can be used to open warded bit-key locks. *a.* True *b.* False

6. If possible, house numbers should be visible from the street. *a.* True *b.* False

7. A standard electromagnetic lock includes a rectangular electromagnet and a rectangular wood and glass strike plate. *a.* True *b.* False

8. A blank is a key that fits two or more locks. *a.* True *b.* False

9. One difference between a bit key and barrel key is that the bit key has a hollow shank. *a.* True *b.* False

10. A key-in-knob lock typically is used to secure windows. *a.* True *b.* False

11. The Egyptians are credited with inventing the first lock to be based on the locking principle of today's pin-tumbler lock. *a.* True *b.* False

12. A jimmy-proof deadlock typically is the most secure type of lock for sliding glass doors. *a.* True *b.* False

13. Lock picking is the most common way homes are burglarized. *a.* True *b.* False

14. Lock impressioning is the most common way homes are burglarized. *a.* True *b.* False

15. A long-reach tool and wedge are commonly used to open locked automobiles. *a.* True *b.* False

16. It's legal for locksmiths to duplicate a U.S. post office box key at the request of the box renter—if the box renter shows a current passport or driver's license. *a.* True *b.* False

17. The Romans are credited with inventing the pin-tumbler lock. *a.* True *b.* False

18. Five common keyway groove shapes are left angle, right angle, square, V, and round. *a.* True *b.* False

19. To pick open a pin-tumbler cylinder, you usually need a pick and a torque wrench. *a.* True *b.* False

20. A Door Reinforcer makes a lock harder to pick open. *a.* True *b.* False

21. Common door lock backsets include
    a. $2^1/_2$ inch and $2^3/_5$ inch.
    b. 3 inch and $2^3/_4$ inch.
    c. $1^3/_4$ inch and $2^1/_2$ inch.
    d. $2^3/_8$ inch and $2^3/_4$ inch.

22. How many sets of pin tumblers are in a typical pin-tumbler house door lock?
    a. 3 or 4
    b. 5 or 6
    c. 11 or 12
    d. 7 or 8

23. Which lock is unpickable?
    a. A Medeco biaxial deadbolt
    b. A Grade 2 Titan
    c. The Club steering wheel lock
    d. None of the above

24. Which are basic parts of a standard key-cutting machine?
    a. A pair of vises, a key stop, and a grinding stylus
    b. Two cutter wheels, a pair of vises, and a key shaper
    c. A pair of vises, a key stylus, and a cutter wheel:
    d. A pair of styluses, a cutter wheel, and a key shaper

25. What are two critical dimensions for code-cutting cylinder keys?
    a. Spacing and depth
    b. Bow size and blade thickness
    c. Blade width and keyhole radius
    d. Shoulder width and bow size

26. Which manufacturer is best known for its low-cost residential key-in-knob locks?

    a. Kwikset Corporation
    b. Medeco Security Locks
    c. The Key-in-Knob Corporation
    d. ASSA

27. The most popular mechanical lock brands in the United States include

    a. Yale, Master, Corby, and Gardall.
    b. Yale, Kwikset, Master, and TuffLock.
    c. Master, Weiser, Kwikset, and Schlage.
    d. Master, Corby, Gardall, and Tufflock.

28. A mechanical lock that is operated mainly by a pin-tumbler cylinder is commonly called a

    a. disk-tumbler pinned lock.
    b. cylinder-pin lock.
    c. mechanical cylinder-pin lock.
    d. pin-tumbler cylinder lock.

29. Burglars target garage doors because

    a. people keep property that is easy to fence in garages.
    b. a garage that's attached to the house provides a discreet way to break into the house.
    c. a garage door with thin or loose panels can be accessed without opening the door.
    d. all the above.

30. Glass is a deterrent to burglars because

    a. it slows a burglar down.
    b. broken shards of glass can injure a burglar.
    c. shattering glass is noisy and attracts attention.
    d. all the above.

31. If you don't feel secure about glass in a window, you can increase security by

    a. replacing the glass with carbonated glass or antihammer plastic.
    b. replacing the glass with impact-resistant acrylic or polycarbonate or high-security glass.
    c. covering the glass with bullet-proof paint.
    d. all the above.

32. Warded bit key locks

    a. provide high security.
    b. provide little security.
    c. are hard to open without the right key.
    d. are usually the best choice for use on an exterior door.

33. Which of the following key combinations provides the most security?

    a. 55555
    b. 33333

    c. 243535

    d. 35353

34. Which of the following key combinations provides the least security?

    a. 243535

    b. 1111

    c. 321231

    d. 22222

35. A blank is basically just

    a. a change key with cuts on one side only.

    b. an uncut or uncombinated key.

    c. any key with no words or numbers on the bow.

    d. a master key with no words or numbers on the bow.

36. You often can determine the number of pin stacks or tumblers in a cylinder by

    a. its key-blade length.

    b. its key-blade thickness.

    c. the key-blank manufacturer's name on the bow.

    d. the material of the key.

37. Spool and mushroom pins

    a. make keys easier to duplicate.

    b. can hinder normal picking attempts.

    c. make a lock easier to pick.

    d. make keys harder to duplicate.

38. As a general rule, General Motors' 10-cut wafer sidebar locks have

    a. a sum total of cut depths that must equal an even number.

    b. up to four of the same depth cut in the 7, 8, 9, and 10 spaces.

    c. a maximum of five number 1 depths in a code combination.

    d. at least one 4-1 or 1-4 adjacent cuts.

39. When drilling open a standard pin-tumbler cylinder, position the drill bit

    a. at the first letter of the cylinder.

    b. at the shear line in alignment with the top and bottom pins.

    c. directly below the bottom pins.

    d. directly above the top pins.

40. When viewed from the exterior side, a door that opens inward and has hinges on the right side is a

    a. left-hand door.

    b. right-hand door.

    c. left-hand reverse-bevel door.

    d. right-hand reverse-bevel door.

41. A utility patent

    a. relates to a product's appearance, is granted for 14 years, and is renewable.

    b. relates to a product's function, is granted for 17 years, and is nonrenewable.

      c. relates to a product's appearance, is granted for 17 years, and is renewable.

      d. relates to a product's function, is granted for 35 years, and is nonrenewable.

42. To earn a UL-437 rating, a sample lock must

      a. pass a performance test.

      b. use a patented key.

      c. use hardened-steel mounting screws and mushroom and spool pins.

      d. pass an attack test using common hand and electric tools such as drills, saw blades, puller mechanisms, and picking tools.

43. Tumblers are

      a. small metal objects that protrude from a lock's cam to operate the bolt.

      b. fixed projections on a lock's case.

      c. small pins, usually made of metal, that move within a lock's case to prevent unauthorized keys from entering the keyhole.

      d. small objects, usually made of metal, that move within a lock cylinder in ways that obstruct a lock's operation until an authorized key or combination moves them into alignment.

44. Electric switch locks

      a. are mechanical locks that have been modified to operate with battery power.

      b. complete and break an electric current when an authorized key is inserted and turned.

      c. are installed in metal doors to give electric shocks to intruders.

      d. are mechanical locks that have been modified to operate with alternating-current (ac) electricity instead of with a key.

45. A popular type of lock used on GM cars is

      a. a Medeco pin tumbler.

      b. an automotive bit key.

      c. a sidebar wafer.

      d. an automotive tubular key.

46. When cutting a lever-tumbler key by hand, the first cut should be the

      a. lever cut.

      b. stop cut.

      c. throat cut.

      d. tip cut.

47. How many possible key changes does a typical disk-tumbler lock have?

      a. 1,500

      b. 125

      c. A trillion

      d. 25

48. Which manufacturer is best known for its interchangeable core locks?

      a. Best Lock

      b. Kwikset Corporation

       c. ILCO Interchangeable Core Corporation
       d. Interchangeable Core Corporation

49. James Sargent is famous for

       a. inventing the Sargent key-in-knob lock.
       b. inventing the time lock for banks.
       c. inventing the double-acting lever-tumbler lock.
       d. being the first person to pick open a Medeco biaxial cylinder.

50. Which are common parts of a combination padlock?

       a. Shackle, case, bolt
       b. Spacer washer, top pins, cylinder housing
       c. Back cover plate, case, bottom pins
       d. Wheel pack base plate, wheel pack spring, top and bottom pins

51. General Motors' ignition lock codes generally can be found

       a. on the ignition lock.
       b. on the passenger-side door.
       c. below the Vehicle Identification Number (VIN) on the vehicle's engine.
       d. under the vehicle's brake pedal.

52. Which code series is used commonly on Chrysler door and ignition locks?

       a. EP 1-3000
       b. CHR 1-5000
       c. CRY 1-4000
       d. GM 001-6000

53. How many styles of lock pawls does General Motors use in its various car lines?

       a. One
       b. Five
       c. Over 5
       d. Three

54. The double-sided (or 10-cut) Ford key

       a. has five cuts on each side; one side operates the trunk and door, whereas the other side operates the ignition.
       b. has five cuts on each side; either side can operate all locks of a car.
       c. has 10 cuts on each side; one side operates the trunk and door, whereas the other side operates only the ignition.
       d. has 10 cuts on each side.

55. Usually the simplest way to change the combination of a double-bitted cam lock is to

       a. rearrange the positions of two or more tumblers.
       b. remove two tumblers and replace them with new tumblers.
       c. remove the tumbler assembly and replace it with a new one.
       d. connect a new tumbler assembly to the existing one.

56. When shimming a pin-tumbler cylinder open

       a. use the key to insert the shim into the keyway.
       b. insert the shim into the keyway without the key.

    c. insert the shim along the left side of the cylinder housing.

    d. insert the shim between the plug and cylinder housing between the top and bottom pins.

57. A lock is any

    a. barrier or closure that restricts entry.

    b. fastening device that allows a person to open and close a door, window, cabinet, drawer, or gate.

    c. device that incorporates a bolt, cam, shackle, or switch to secure an object—such as a door, drawer, or machine—to a closed, locked, on, or off position and that provides a restricted means—such as a key or combination—of releasing the object from that position.

    d. device or object that restricts entry to a given premise.

58. Which wheel in a safe lock is closest to the dial?

    a. Wheel 1

    b. Wheel 2

    c. Wheel 3

    d. Wheel 0

59. Which type of safe is best for protecting paper documents from fire?

    a. Paper safe

    b. Money safe

    c. UL-listed record safe

    d. Patented burglary safe

60. Which type of cylinder is typically found on an interlocking deadbolt (or jimmy-proof deadlock)?

    a. Bit-key cylinder

    b. Key-in-knob cylinder

    c. Rim cylinder

    d. Tubular deadbolt cylinder

Registered Security Professional Answer Sheet

Make a photocopy of this answer sheet to mark your answers on.

1. ____	21. ____	41. ____
2. ____	22. ____	42. ____
3. ____	23. ____	43. ____
4. ____	24. ____	44. ____
5. ____	25. ____	45. ____
6. ____	26. ____	46. ____
7. ____	27. ____	47. ____
8. ____	28. ____	48. ____
9. ____	29. ____	49. ____
10. ____	30. ____	50. ____

11. ____	31. ____	51. ____
12. ____	32. ____	52. ____
13. ____	33. ____	53. ____
14. ____	34. ____	54. ____
15. ____	35. ____	55. ____
16. ____	36. ____	56. ____
17. ____	37. ____	57. ____
18. ____	38. ____	58. ____
19. ____	39. ____	59. ____
20. ____	40. ____	60. ____

To receive your Registered Security Professional certificate, just submit your answers to this test (a passing score is 70 percent) and enclose a check or money order for US$50 (nonrefundable) payable to IAHSSP. And enclose copies of any two of the following items (do not send original documents because they will not be returned):

- City or state locksmith license
- Driver's license or state-issued identification
- Locksmith suppliers invoice
- Certificate from locksmithing or security school or program
- Yellow Pages listing
- Business card or letterhead from your company
- Association membership card or certificate
- Locksmithing, safe technician, or alarm installer bond card or certificate
- Letter from your employer or supervisor on company letterhead stating that you work as a locksmith or security professional
- Letter of recommendation from a Registered Professional Locksmith or Registered Security Professional
- Copy of an article you've had published in a locksmithing or security trade journal
- ISBN number and title of locksmith or security-related book you wrote

Send everything to IAHSSP, Box 2044, Erie, PA 16512-2044. Please allow 6 to 8 weeks for processing.

Name: _____ Title:

_____

Company name:

_____

____

Address:

_____
_____

City, State, Zip
Code:_____

Telephone number: _____ E-mail address:

_____

Installing a Securitron Mortise Glass Lock

INSTALLATION AND OPERATING INSTRUCTIONS
FOR
Mortise Glass Lock MGL-12/24 RH/LH

1. INTRODUCTION

Securitron's MGL-12/24 is an electromechanical "Fail Secure" mortise lock designed for use in aluminum framed glass doors (prepped for Adams Rite MS Deadbolt). This unit's slim design provides for mounting into a wide range of narrow stile doors and aluminum door frame profiles. The unit is available in 12VDC or 24VDC and can be obtained from the factory in either Right Hand or Left Hand configuration. The door can be opened when the lock is powered. Free egress is provided via the included door handle. A mechanical key cylinder override required for the opposite side of the door is not included. The MGL comes standard with both lock and handle status monitoring for easy interface with any access control system.

2. SPECIFICATIONS

Lock Housing Dimensions:
　　Inches: 1"W X 7"H X 1-7/8"D
　　Millimeters: 25 W X 178 H X 48 D

Electrical:
　　0.40A at 12VDC
　　0.20A at 24VDC

3. PRODUCT OVERVIEW

Upon unpacking this product, an inventory should be made to ensure that all the required components and hardware have been included. Along with these Instructions this product should include the following items:

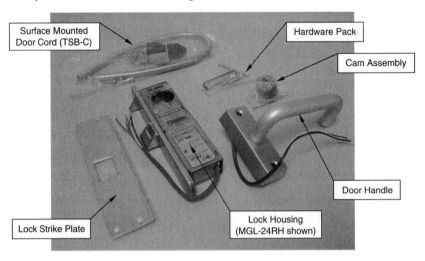

Surface Mounted Door Cord (TSB-C)

Hardware Pack

Cam Assembly

Door Handle

Lock Housing (MGL-24RH shown)

Lock Strike Plate

An ASSA ABLOY Group company   ASSA ABLOY

4. RECOMMENDED TOOLS

Standard and Phillips Screwdrivers
Wire Strippers
Wire Crimping Tool (and Connectors)
Drill Motor
Drill bits: 1/8", #21(For 10-32 UNF Tap), 3/16", 5/16"
Thread Tap, 10-32UNF (Bottom Tap)

Hex (Allen) Wrenches: 5/64" (Provided)
Hammer, Small (Ball Peen)
Jig Saw (with Blade for cutting Aluminum)
Hole Saw Bits: 1-1/8", 1-1/4"
Router (Optional) with 3/16" Max. Dia. Bit

5. INSTALLATION INSTRUCTIONS

5.1. Pre-Installation Survey

It is recommended that an initial survey be made of the area where the lock will be installed. Determination of optimal mounting placement should be made prior to the installation with considerations for the following:

A) The structural integrity of the mounting surfaces (i.e. door and frame) must be strong enough to meet the requirements of the lock. The mounting area must be of sufficient dimension that the cutting required in the installation will not weaken the structural integrity of the door or frame.

B) The lock and its wiring must be protected to a reasonable degree from potential damage due to intruders or vandals.

Note: Sections 5.2 and 5.3 provide instruction for the installation of a standard right hand lock (right hand door). For general information on lock/door handing see Section 7 Appendix B. For changing the existing handing of the MGL-12/24 door handle see Section 7 Appendix C. If changing the handing of the MGL-12/24 lock housing is necessary in the field, please contact the factory at 1-800-MAGLOCK for detailed instructions.

5.2. Lock Housing Installation

5.2.1. Door Preparation

The following illustration (Figure 1) displays the preparation required to install the MGL lock housing to an aluminum framed glass door:

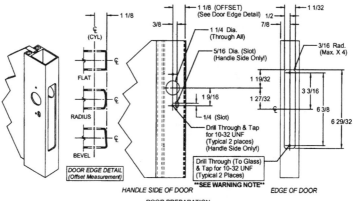

DOOR PREPARATION
Figure 1

WARNING: Care must be taken when drilling (and tapping) these two (2) holes. DO NOT allow drill bit or tap to come into contact with glass!

5.2.2. Installation of Lock Housing in Door

This section will provide step-by-step instructions for the installation of the lock housing into an aluminum framed glass door.

The following illustration (Figure 2) is an exploded view of the assembly of a lock housing into a glass door frame:

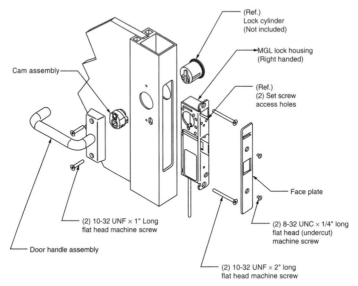

Figure 2

Installation:

1. Prepare door as shown in Figure 1.
2. Run electrical wiring as required through door and frame. (See Section 5.4. for installation of the surface mounted door cord TSB-C).
3. Remove face plate from the lock housing by removing the two (2) 8-32 UNC X 1/4" long flat head (undercut) machine screws. Set parts aside for later reinstallation.
4. Retrieve wiring from inside door (through slot cut in edge of door) and using wire strippers, crimpers and connectors, connect to lock housing wiring (5-wire cable) as required (see electrical wiring diagram in Section 6.1 for proper connections).
5. If the door handle electronic functions are required, insert the (3-wire cable) through the door (3/16" Slot) and connect to wiring as required (see electrical wiring diagram in Section 6.1 for proper connections). If these functions are not required the micro switch in the handle may be removed or the handle may be installed without connecting the wires.
6. Insert the lock housing into the slot opening in the edge of the door inserting the wire end first.
7. After aligning the countersunk holes in the ends of the lock housing face with the tapped holes in the door (behind slot), insert the two (2) 10-32 UNF X 2" long flat head machine screws through the lock housing and loosely thread into the tapped holes. (**DO NOT** completely tighten at this point). **Note:** As necessary, tuck wiring out of the way of the installation hardware (the lock electrical cable tends to interfere with bottom lock mounting screw).
8. Install any dress bezel/hardware as required and insert the threaded end of the lock cylinder (customer provided) through 1-1/4" diameter hole in door and thread into lock housing. Ensure the actuating lever of the lock cylinder engages the dead latch armature linkage of the lock (internal) with the key receptacle (slot) of the cylinder in the down position.

9. Insert the lever end of the cam assembly through the opposite 1-1/4" diameter hole in door and into the lock housing with the square (cam drive) hole in the down position.

10. Using a 5/64" hex (Allen) wrench, tighten the two (2) set screws through the access holes in the face of the lock housing (see Figure 2) to secure the lock cylinder and the cam assembly in place.

11. Insert the square drive shaft of the door handle into the square (cam drive) hole of the cam assembly. Align the mounting holes of the door handle with tapped holes in the door, then install the two (2) 10-32 UNF X 1" long flat head machine screws and tighten using a screwdriver.

12. Using a screwdriver, tighten the two (2) 10-32 UNF X 2" long flat head machine screws to secure the lock housing into position.

13. Test the handle actuation. If necessary, loosen screws as required, and adjust the assembly so that it functions smoothly.

14. Replace the face plate to the lock housing using a screwdriver and the two (2) 8-32 UNC X 1/4" long flat head (undercut) machine screws removed in Step 3.

5.3. Strike Plate Installation

Note: A strike plate has been provided and is considered necessary for wear protection of the door frame and for the proper function of the lock mechanism.

5.3.1. Door Frame Preparation

The following illustration (Figure 3) displays the preparation required to the door frame for the proper installation of the strike plate when an MGL is assembled in an aluminum framed glass door:

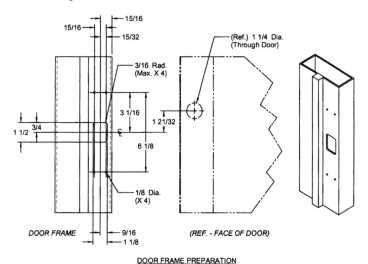

DOOR FRAME PREPARATION

Figure 3

5.3.2. Installation of Strike Plate on Door Frame

This section will provide step-by-step instructions for the installation of the lock strike plate onto an aluminum door frame.

The following illustration (Figure 4) is an exploded view of the assembly of a strike plate onto the door frame:

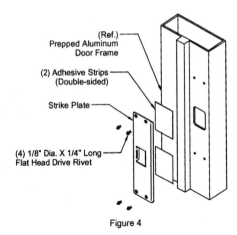

Figure 4

Installation:

1. Prepare door frame as shown in Figure 3. **Note:** The strike plate may be used as a template for accurately locating the mounting holes to be drilled in the frame.
2. Remove the strike plate and double-sided adhesive strips from the packaging.
3. Remove one (1) side each of the adhesive strips and apply each to the back side of the strike plate (one on each end of the rectangular cut-out at center). **Note:** These adhesive strips **must** be used to provide proper strike support.
4. Remove the remaining side from each of the adhesive strips, then position the strike plate (by aligning the holes in the plate with the holes in the frame) and affix the strike plate to the door frame.
5. Insert the four (4) drive rivets into the holes of the strike plate so that the head of the rivet is flush with the face of the strike plate.
6. Using a small hammer, strike the drive pin of one of the rivets until the pin is seated flush with the head of the rivet. Repeat for the next three (3) rivets to secure the strike plate.
7. If necessary, to provide a smooth closing operation between the latch and the strike plate, a small amount of silicone based grease may be applied to the strike and/or latch plungers.

5.4. Surface Mounted Door Cord (TSB-C) Installation

The following illustration (Figure 5) is an exploded view of a typical installation of the door cord for the routing of the electrical wiring to the lock:

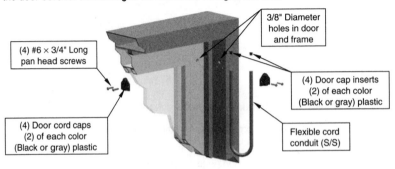

Figure 5

Installation:

1. Drill a 3/8" diameter hole in door and frame (hinge side) as shown in Figure 5.
2. Insert one (1) door cap Insert into each end of the flexible cord conduit.
3. Run electrical wiring as required through the flexible cord conduit (with inserts) and holes in door and frame.
4. Use door cord cap as a template to mark/drill 1/8" pilot holes for mounting.
5. Insert each end of the conduit into the recess of each of the cord caps and mount each cap to the door and frame using a screwdriver and the two (2) each #6 X 3/4" long (Phillips) pan head screws provided.

6. OPERATIONAL INSTRUCTIONS

Once wired, as shown in the wiring diagram (Figure 6) below, the MGL will operate in the following manner:

At rest the door is locked from outside of the secured area. The handle allows free and immediate egress from inside the secured area. When the normally opened switch contact is activated the lock releases and allows access from outside of the secured area. The lock monitoring allows the condition of the device to be monitored by access control or intrusion detection systems. The monitoring switch in the handle will send a REX (request to exit) input to the access control or intrusion detection system allowing free egress in normal conditions without door forced or trouble alarms.

6.1. ELECTRICAL WIRING

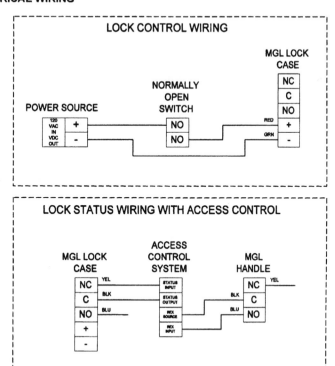

Figure 6

6.1.1. WIRE GAUGE SIZING

If the power supply is at a distance from the device, voltage will be lost (dropped) in the connecting wires so that the unit will not receive full voltage. The following chart shows the minimum wire gauge that will hold voltage drop to an acceptable 5% for different device to power supply distances. Proper use of the chart assumes a dedicated pair of wires to power each unit (no common negative). Note that a device operating on 24 volts is a much better choice for long wire runs as it has 4 times the resistance of a 12 volt installation.

Also note that the correct calculation of wire sizing is a very important issue as the installer is responsible to insure that adequate voltage is supplied to any load. In multiple unit installations, the calculation can become quite complex so refer to Section 7 Appendix A for a more complete discussion.

Distance	Gauge 12V	Gauge 24V	Distance	Gauge 12V	Gauge 24V
80 FT	20 GA	24 GA	800 FT	10 GA	16 GA
200 FT	17 GA	22 GA	1500 FT	8 GA	14 GA
400 FT	14 GA	20 GA	3000 FT	N/A	12 GA

7. APPENDICES

A. CALCULATING WIRE GAUGE SIZING

The general practice of wire sizing in a DC circuit is to avoid causing voltage drops in connecting wires that reduce the voltage available to operate the device. As the MGL-12/24 is a very low power device, it can be operated long distances from its power source. For any job that includes long wire runs, the installer must be able to calculate the correct gauge of wire to avoid excessive voltage drops.

This is done by taking the current draw of the lock and multiplying by the resistance of the wire I x R = Voltage drop (i.e. 0.100A x 10.1 Ohms = 1.01 Volts dropped across the wire). For all intents and purposes it can be said that a 5% drop in voltage is acceptable so if this were a 24 Volt system (24 Volts x .05 = 1.2 Volts) a 1.01 Volt drop would be within tolerance.

To calculate the wire resistance, you need to know the round trip distance from the power supply to the MGL-12/24 and the gauge (thickness) of the wire. The following chart shows wire resistance per 1000 ft (305 meters):

Wire Gauge	Resistance/1,000 ft	Wire Gauge	Resistance/1,000 ft
8 Gauge	.6 Ohms	16 Gauge	4.1 Ohms
10 Gauge	1.0 Ohms	18 Gauge	6.4 Ohms
12 Gauge	1.6 Ohms	20 Gauge	10.1 Ohms
14 Gauge	2.5 Ohms	22 Gauge	16.0 Ohms

B. LOCK / DOOR HANDING – General Information

The following illustration (Figure B-1) shows the various handing configurations for standard and reverse mortise lock applications:

STANDARD LOCKS
In Standard Locks, the FLAT Side of the Latch Bolt faces TOWARD
the Handle Side of the Lock (AWAY from The Lock Key Cylinder)

Left Hand
(Left Hand Door)   KEY CYLINDER SIDE   Right Hand
(Right Hand Door)

REVERSE LOCKS
In Reverse Locks, the FLAT Side of the Latch Bolt faces AWAY
from the Handle Side of the Lock (TOWARD The Lock Key Cylinder)

Left Hand Reverse
(Right Hand Door)   KEY CYLINDER SIDE   Right Hand Reverse
(Left Hand Door)

Figure B-1

Note: Reverse options may be achieved simply by reversing the positions of the lock cylinder and the door handle assembly.

C. HANDING FOR THE MGL-12/24 DOOR HANDLE

This section will provide step-by-step instructions for reversing the handing of the door handle provided with the MGL.

It should be noted that the figures (photographs) used in this instruction show the changeover of a door handle provided with right hand (RH) version of the MGL. Therefore if a changeover of the door handle provided with the left hand (LH) version is to be performed, reverse orientation to those shown here will apply.

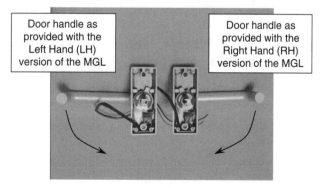

Door handle as provided with the Left Hand (LH) version of the MGL

Door handle as provided with the Right Hand (RH) version of the MGL

• **Tools required for this procedure** (minimum):

Flat Blade Screwdriver
Adjustable Tongue & Groove Pliers (channel lock)
Needle-Nose Pliers (Optional)

Procedure:

1) Remove door handle from packaging box and place on clean, dry work surface.

2) Using a flat blade screwdriver remove the E-clip snap ring, then the handle actuation washer as shown in **Figure C-1** and **Figure C-2**.

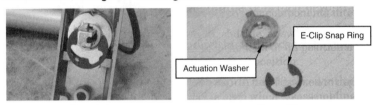

Figure C-1   Figure C-2

3) Remove the torsion spring shown in **Figure C-3**, rotate handle in the assembly 180 degrees, flip spring over and reinstall over handle shaft as shown in **Figure C-4**.

Figure C-3   Figure C-4

4) Flip the handle actuation washer over (alignment key tabs facing up) and reinstall as shown in **Figure C-5**. Using adjustable tongue and groove (channel lock) pliers install E-clip as shown in **Figure C-6**.

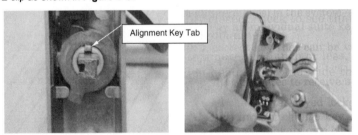

Figure C-5   Figure C-6

5) Handle should now be reversed from original handing as shown in **Figure C-7**.

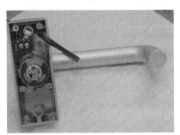

Figure C-7

8. WARRANTY

MAGNACARE® LIMITED LIFETIME WARRANTY

SECURITRON MAGNALOCK CORPORATION warrants that it will replace at customer's request, at any time for any reason, products manufactured and branded by SECURITRON.

SECURITRON will use its best efforts to ship a replacement product by next day air freight at no cost to the customer within 24 hours of SECURITRON's receipt of the product from customer. If the customer has an account with SECURITRON or a valid credit card, the customer may order an advance replacement product, whereby SECURITRON will charge the customer's account for the price of the product plus next day air freight, and will credit back to the customer the full amount of the charge, including outbound freight, upon SECURITRON's receipt of the original product from the customer.

SECURITRON's sole and exclusive liability, and customer's sole remedy, is limited to the replacement of the SECURITRON product when delivered to SECURITRON's facility (freight and insurance charges prepaid by customer). The replacement, at SECURITRON's sole option, may be the identical item or a newer unit which serves as a functional replacement. In the event that the product type has become obsolete in SECURITRON's product line, this MAGNACARE warranty will not apply. This MAGNACARE warranty also does not apply to custom, built to order, or non-catalog items, items made by others (such as batteries), returns for payment, distributor stock reductions, returns seeking replacement with anything other than the identical product, or products installed outside of the United States or Canada. This MAGNACARE warranty also does not apply to removal or installation costs.

SECURITRON will not be liable to the purchaser, the customer or anyone else for incidental or consequential damages arising from any defect in, or malfunction of, its products. SECURITRON does not assume any responsibility for damage or injury to person or property due to improper care, storage, handling, abuse, misuse, or an act of God.

EXCEPT AS STATED ABOVE, SECURITRON MAKES NO WARRANTIES, EITHER EXPRESS OR IMPLIED, AS TO ANY MATTER WHATSOEVER, INCLUDING WITHOUT LIMITATION THE CONDITION OF ITS PRODUCTS, THEIR MERCHANTABILITY OR FITNESS FOR ANY PARTICULAR PURPOSE.

D

Finish Codes

Two major organizations specify architectural finish codes: Builders Hardware Manufacturers Association (BHMA) and American National Standards Institute (ANSI). ANSI codes are often referred to as "U.S. codes."

ANSI and BHMA codes refer to information such as the base metal, what the product is made of, and the color or finish. The color and the finish are two different things. Color is made by applying different shades of pigmented material (such as paint and lacquer) over a base material. Finish refers to the last step in the manufacturing process (such as brushing, rubbing, or antiquing).

Letter Finish Codes

A = Mill Finish Aluminum

AB = Antique Brass

ABA = Antique Brass Aluminum

ABP = Antique Brass Plated

AL = Aluminum

B = Brass

BLK = Black ousing/Chrome Lever (Kaba PowerLever Series)

BN = Satin Bronze

BP = Brass Plated

BPA = Brass Plated Aluminum

BZP = Bronze Plate

C = Chrome

CP = Chrome Plated

D = Dark Bronze Anodized Finish

DL = Duronotic

DU = Duronotic

PB = Polished Brass (equal to 3/US3)

PC = Primed and ready for paint

S = Stainless Steel (equal to 32D)

SB = Satin Brass

SL = Silver Coated

ANSI, U.S., and Other Codes

3 & US3 = Polished Brass, Clear Coated

4 & US4 = Satin Brass, ClearCoated

5, 5A & US5 = Antique Brass, Clear Coated

US9 = Bright Bronze

10 = Bronze

US10 = Satin Bronze, Clear Coated

US10A = Antique Bronze

10B & US10B = Oil Rubbed Bronze

US14 = Bright Nickel Plated

US15 = Satin Nickel Plated

US15A = Antique Nickel

US17A = Blackened Nickel, Clear Coated

26 & US26 = Bright Chrome

26D & US26D = Satin Chrome, Brushed; a.k.a. 'Dull Chrome'

27 = Aluminum

28 & US28 = Satin Aluminum, Clear Anodized; a.k.a. 'Aluminum'

US32 = Bright Stainless Steel

32D & US32D = Satin Stainless Steel; a.k.a. 'Dull Stainless Steel'

BHMA Codes

313 = Duronotic

326 = Brass/Chrome split finish

332 = Polished Brass/Stainless Steel Split Finish

600 = Prime Coated

605 = Polished Brass, Clear Coated

606 = Satin Brass, Clear Coated

609 = Antique Brass, Clear Coated

611 = Bright Bronze

612 = Satin Bronze, Clear Coated

616 = Antique Bronze

613 = Oil Rubbed Bronze

618 = Bright Nickel Plated

619 = Satin Nickel Plated

620 = Antique Nickel

621 = Blackened Nickel, Clear Coated

624 = Duronotic

625 = Bright Chrome

626 = Satin Chrome, Brushed; a.k.a. Dull Chrome

628 = Satin Aluminum, Clear Anodized

629 = Bright Stainless Steel

630 = Satin Stainless Steel

689 = Painted Silver Aluminum

707 = Bright Brass Anodized Aluminum

Trade Journals

The Canadian Locksmith
Arnold Sintnicolass, Publisher
137 Vaughn Road
Toronto, Ontario M6C2L9
Canada
Phone: 905-294-0660

HS Today (free)
P.O. Box 9789
McLean, VA 22102-9789
Phone: 800-503-6506
Web site: www.HSToday.us

The Institutional Locksmith (free)
Greg Mango, Editor
1533 Burgundy Parkway
Streamwood, IL 60107-1861
Phone: 630-837-2044

Institutional Locksmith Association
P.O. Box 24772
Philadelphia, PA 19111

Keynotes (trade journal of Associated Locksmiths of America)
Jim DeSimone, Editor
3003 Live Oak Street
Dallas, TX 75204
Phone: 800-532-2562

Locksmith Ledger International
Gale Johnson, Editor
3030 Salt Creek Lane, Suite 200
Arlington Heights, IL 60005

The National Locksmith
Greg Mango, Editor
1533 Burgundy Parkway
Streamwood, IL 60107-1861
Phone: 630-837-2044

SAVTA (trade journal of Safe and Vault Association)
3500 Easy Street
Dallas, TX 75247
Phone: 214-819-9771

SDM (free)
Laura Stepanek, Editor
1050 IL Route 83, Suite 200
Bensenville, IL 60106-1096
Phone: 630-616-0200

Security Dealer (free)
Susan Brady, Editor-in-Chief
1849 N.E. Acapulco Drive, Suite 16
Jenson Beach, FL 34957

Security Magazine (free)
Bill Zalud, editor
1050 IL Route 83
Bensenvile, IL 60101

Security Systems News (free)
P.O. Box 998
106 Lafayette Street
Yarmouth, ME 04096
Phone: 207-846-0600
Fax: 207-846-0657

Installing Securitron Magnalock Devices*

---

*Courtesy of Securitron Magnalock Corp.

SECURITRON MODEL DK-11W DATA OUTPUT DIGITAL KEYPAD
INSTALLATION & OPERATING INSTRUCTIONS

1. DESCRIPTION

Securitron's DK-11W is a one piece digital keypad designed to output **Weigand 2601 format data** and therefore integrate into an access control system just as if it was a card reader. It is furnished on a single gang, stainless steel outlet box cover with two plastic backboxes respectively for flush or surface mounting. Two LED's (green and red) controlled by the system are furnished.

The DK-11W is not waterproof and is intended for **interior use**. However, an accessory cover (part # WCC) allows its use outside where direct, heavy rain is not expected.

2. PHYSICAL INSTALLATION

Note that the DK-11W is supplied with a choice of tamper resistant #6 spanner machine screws and conventional screws for attaching the plate to the backbox.

Two backboxes are supplied with the DK-11W. The blue backbox comes with a template and is used for **flush mounting** on dry wall or other material where a cut-out can be made. The beige two-piece backbox allows **surface mounting** on a variety of materials. To use the beige backbox, note that its cover and base are snapped together and must first be separated by either pulling the outer rim of the cover away from the base or inserting a screwdriver into the four holes at the corners of the cover and prying the base loose. Once the base is separated from the cover, remove the large rectangular knockout in the center of the base by cutting around it with a knife and then popping it out. The base can then be mounted with the supplied #6 sheet metal screws and plastic anchors. The DK-11W mounts on the cover with the supplied #6 machine screws and then the cover with DK-11W snaps into the mounted base. Wires are usually pulled through the center of the base although it is also possible to attach plastic wiremold raceway to the side of the cover (note the knockouts on the inside of the cover sides).

The DK-11W can be used outdoors with the optional rain cover (part #WCC) although we do not advise this use in areas exposed to heavy, direct rain. When used outdoors, you must supply a weatherproof, gasketed backbox (available from Securitron under part #WBB).

3. WIRING

3.1 POWER AND DATA WIRING

Figure one shows the rear of the DK-11W. You will make connections to the six terminals as shown in the drawing and either leave the jumper block in the factory set position (connects pins 2 and 3) if you plan to power the DK-11W with 12 VDC or move the jumper to connect pins 1 and 2 if you will be using 5 VDC. Note that **operation at 12 volts** with the jumper block in the 5 volt position **can damage the unit**.

Note that **the DK-11W will not operate on AC power**. It will, however, accept **full wave rectified DC power** (transformer + bridge rectifier) **when it is being powered by 12 VDC**. When it is being powered by **5 VDC, the voltage must be regulated** (+/- 1/2 volt). Be sure to **observe polarity** when you power the DK-11W.

FIG. 1: CIRCUIT BOARD OVERVIEW

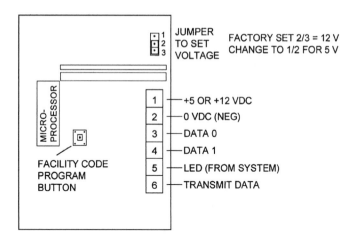

The DK-11W will draw a maximum of **35 mA at 5 VDC or 12 VDC**.

Note that space does not permit numbering the six terminals on the circuit board. When making your connections, you have to refer to the drawing and mentally count off the terminals.

The Weigand output terminals: Data 0 and Data 1 connect to the appropriate inputs of the access control system. The **wire run maximum distance** for reliable operation depends on the wire gauge. A guide line is 200 ft. for 22 gauge; 300 ft. for 20 gauge and 500 ft. for 18 gauge.

3.2 LED AND "TRANSMIT DATA" WIRING

The LED's on the DK-11W follow the convention for card readers. When a "high" signal (+5 VDC) is connected to the LED terminal, the red LED will be on and the green LED will be off. When this input goes "low" (0 VDC), the green LED will be on and the red LED will be off. This flipping of the LED's is controlled by the access system and typically prompts the user when his entry has been accepted (or not accepted).

The "transmit data" terminal is not used with most systems. When it is unconnected, the code sequence is automatically transmitted to the system following key entry (see Section 5). However, with some systems, the controller can be busy and must therefore remotely command data transmission. If this input is in a low state (connected to 0 VDC), the DK-11W will store the code sequence until the input goes high (receives +5 VDC). The code will then be transmitted as the system will be prepared to receive the sequence and release the door. Naturally, while a code is being stored, the keypad will ignore further inputs as the delay prior to the system commanding transmission of the code sequence will be very brief.

4. FACILITY (SITE) CODE PROGRAMMING

In the Weigand 26 bit code format (also called 2601), the first eight active bits constitute a facility or site code. These eight bits correspond to standard numbers 0-254. The access control system normally expects to see a "two part" transmission wherein the eight bit facility code precedes the 16 bit PIN code which identifies the individual who is requesting entry. Possible PIN codes convert to standard numbers 0-65,534. The reason for the creation of the facility code is to enhance card security as on a card, both the facility and PIN codes are stored. If a card was transported to a different facility, it would not be accepted by the different system even though the PIN code happened to be valid because the facility code would not be.

With a digital keypad like the DK-11W, the facility code required by the system must be internally stored since a person requesting entry will only know his PIN code. The DK-11W ships with a **factory set facility code of "0"**. To **change the facility code** to the one in use by the system, identify the program button on the unit's circuit board (see Figure 1). With the unit powered, press the button until you hear a steady beep. This annunciates **facility code program mode**. If you do nothing, the unit will automatically drop out of program mode **in 30 seconds** and the facility code will not be changed. To change the code, during this 30 second window, simply enter the new facility code. You don't have to enter three digits if the facility code is less than 100 (leading zeros are not necessary). Do not pause more then five seconds between digits as the unit has an internal timer that resets five seconds after a key press. After you have completed site code entry, you can press "*" or "#" to terminate the sequence or simply wait five seconds. You will receive a **single beep to confirm a good entry**. If you have entered a sequence that is too large (a number greater than 254), you will receive a **double beep** (error). This is your prompt to re-enter the code. To do this, you will have to press the program button another time as the unit will not remain in program mode after data entry.

The DK-11W employs non-volatile EEPROM memory so that the facility code is **retained in a power failure**.

5. OPERATION

To operate the unit, simply enter the PIN code (from 0-65534) and then either press * or # or wait five seconds. Note that successful key presses are **echoed by a beep**. The PIN code together with the site code prefix stored in the unit will then be sent to the access control system. **Do not pause more than five seconds** between digits or an incomplete sequence will be transmitted as the unit automatically transmits when it does not see any key presses for five seconds. The system will respond by allowing entry or not and will generally annunciate this by control of the two LED's. How the LED's are used exactly will vary from system to system. If you enter a number larger than 65,534, the DK-11W will reject the sequence and transmit nothing. This rejection is communicated by two beeps (the error signal).

APPENDIX A: 2601 CODE STRUCTURE

The 26 bit transmission begins with a parity bit followed by 24 code bits and ended by a second parity bit. The first parity bit is even parity calculated over the first 12 code bits as follows: if the 12 bits sum to 0, the parity bit is set to 0. If the 12 bits sum to 1, the parity bit is set to 1. The second (ending) parity bit is odd parity calculated over the second 12 code bits as follows: if the second 12 bits sum to 0, the parity bit is set to 1. If the second 12 bits sum to 1, the parity bit is set to 0.

The 24 code bits have internal structure as follows. The first eight bits are the facility code. The next 16 bits are the PIN code. All data is transmitted Most Significant Bit first from the keypad. The transmission begins with the even parity bit, proceeds through the eight bit facility code followed by the 16 bit PIN code and ends with the odd parity bit.

The transmission of a 0 bit occurs when the data 0 line transitions below 1.1 V for 50 microseconds. The transmission of a 1 bit occurs when the data 1 line transitions below 1.1 V for 50 microseconds. The interval between bit transmitting pulses is one millisecond.

MAGNACARE® LIMITED LIFETIME WARRANTY

SECURITRON POWER SUPPLIES
MODELS: BPSM-12-6, 9 AND 15; BPSM-24-4, 6 AND 10
OPERATION AND INSTALLATION INSTRUCTIONS

1. DESCRIPTION

These instructions cover 6 different models as shown above. The part number expresses first the **output voltage** (12 or 24V DC) and second the **maximum output current capacity**. For example, a model BPSM-24-6 can supply up to 6 amps at 24 volts. Securitron large power supplies consist of a power module and CCS8M control board to which all installer connections are made. The board accomplishes several functions. It provides terminals for line voltage input and eight separate DC output circuits, so that a number of devices can be powered. Each control circuit has an individual slide switch to turn it on and off and an LED to annunciate its status. The CCS8M control board also provides the following:

- AC input and DC output LED indicators
- Emergency release terminals (fire alarm disconnect)
- AC status (form "C" contact) and DC status (form "C" contact) terminals
- Line voltage and DC fuses
- Built-in charger for sealed lead acid or gel cell batteries
- Automatic switch over to stand-by battery when AC fails

All power supplies in the BPSM series are **Class 2 rated** when installed following these instructions.

2. SAFETY

Two hazards are present in the supply. Line voltage input presents a high voltage shock hazard and the DC/battery output, represent a high energy (current) hazard. A shorted battery can swiftly supply levels of current sufficient to melt wiring insulation and cause a fire. To insure safety, note first that the cover LED (red) is on at any time that the supply is dangerous, which is either if it is receiving line voltage or if batteries are operating. **The supply enclosure must only be opened by trained service personnel when the red LED is on. The green LED on the CCS8M control board is on when AC line voltage is present.** Other safety features include line voltage and DC fuses and the fact that the line voltage input terminals are under a warning guard plate.

An ASSA ABLOY Group company   ASSA ABLOY

3. OPERATING CHARACTERISTICS

3.1 LINE VOLTAGE INPUT

110-120 VAC should be input to terminals "H", "N", "G", as shown in the drawing. This is fed to the input of the power module through factory made connections. The green LED will illuminate when AC line voltage is present. The line voltage current drawn by the power supply module will be approximately half the DC output. For example, for a 4 amp power supply, the line voltage service should be able to supply at least 2 amps. Note: if the suffix "H" appears in the part number (i.e. BPSMH-24-4), the unit requires 220 VAC input. Apart from this change, all other characteristics are the same.

3.2 DC OUTPUT

The maximum DC output of the power supply is expressed by the final figure in the part number. The BPSM-24-4 can supply up to 4 amps; the BPSM-12-6, up to 6 amps etc. However, these power supplies are adjustable and the voltage level set affects the current output capacity. When **used with batteries**, the power supplies must be set at 12.5% over voltage (13.5 V for 12 volt supplies and 27 V for 24 volt supplies). **This is the factory setting** and if the voltage is to be changed, use the potentiometer marked "V Adj" on the power supply module. The current rating takes into account the possibility of the supplies operating at 12.5% over voltage. Therefore any supply which is operated at its nominal voltage (12 or 24 v) can supply about 20% more than its rated current. Despite this, we **strongly recommend** that supplies be operated substantially below their maximum output capability. Operating power supplies at their maximum greatly increases the possibility of heat induced failure. "Margin for error" is lost and this is inappropriate for a security system. **Power supplies should be run at no more than two thirds of their maximum capacity for optimum reliability.**

3.3 EMERGENCY RELEASE (FIRE ALARM) TERMINALS

If the power module is operating or if batteries are operating, the red LED will illuminate. +V (12 or 24VDC depending on the model power supply) will then be on terminal F1. A connection must then be made between terminals F1 and F2 (this will turn on a power relay) before +V is routed to the "P" terminals. **Terminals F1 and F2 therefore constitute an emergency release point.** If desired, for instance, NC contacts controlled by the user's fire alarm system can be connected across terminals F1 and F2 such that the connection between these terminals will be broken in the event of a fire. **UL listed auxiliary latching normally closed contacts (minimum switching capability of 75 mA) from the fire alarm system should be used.** "Trouble" contacts must not be used. This will automatically release all the devices being driven by the unit. If the emergency release terminals are not to be used in this way, a jumper should be placed between them so that the board's output terminals will function.

Terminal FA is a free parking terminal.

3.4 OUTPUT TERMINALS

The CCS8M board has three types of output terminals. **"P" terminals** are on individual circuit breakers and carry +12 or +24 volts on them (when the emergency release terminals are closed). **The "H" terminal** carries the full +V output of the supply on a single terminal (when F1 and F2 are closed). Use the "H" terminal for applications where the device being powered requires more than 2 Amps of current. The Polyswitch circuit breakers cannot reliably supply more than 2 Amps of current without tripping and **you should never wire multiple "P" terminals in parallel** to supply increased current. This bypasses the safety role of the Polyswitch breakers and also does not work very well. When two "P" terminals are wired in parallel, current carrying capacity is not doubled. The current conducted through the two terminals will not be identical so one switch will break first and then the second will immediately trip. When "P" terminals are correctly used as isolated outputs, each is inherently current limited to **Class 2** standards. Always use the "H" terminal for applications requiring high current. Finally, the **"R" terminals** are all for 0 volt DC negative return and are in common.

3.5 SUPERVISION

During normal operation both trouble reporting relays are energized (refer to Voltage Supervision Diagnostic Table on page 4). To report loss of AC, connect corresponding wiring to terminals marked **AC FAIL** (If AC line voltage is present, there will be continuity between the **C and NO terminals**). To report low battery, connect corresponding wiring to terminals marked **LOW BATT.** During battery operation and AC line voltage not present, there will be continuity between the **C and NO terminals**. If during battery operation the battery voltage drops below 10.4VDC (20.8VDC for the 24V power supplies) the battery will automatically disconnect to prevent possible damage to the battery. **Once the battery has been automatically disconnected, AC line voltage must be restored before the battery is again available for backup.** All relay contacts have transient/surge absorbers to protect the board from external high voltage transients. **The maximum voltage allowed on any pair of relay contact terminals is 30VAC or 38VDC.**

3.6 FUSING AND CIRCUIT POLYSWITCHES

An **AC fuse, DC fuse** and eight **Polyswitches** are present on the board. The AC fuse is on the hot 120 VAC input and protects against an internal short in the power supply transformer. A short in the DC load will not blow the AC fuse. The DC fuse protects the full DC output of the supply prior to it being divided through the Polyswitches to the individual "P" outputs. The Polyswitch is a special type of automatic circuit breaker. If one of the Polyswitches receives an overload, it will rapidly cut the current down to a small leakage current (about 100 mA) which will **allow the other outputs to continue to operate**. Note that each "P" output includes a slide switch and LED. The slide switch can cut DC power to its respective output and the LED monitors when the output is powered. In the event of one of the Polyswitches tripping, the associated LED will go out or dim. If all the LED's go out, one of the fuses has tripped or the power supply has gone into automatic shut off (discussed later). **Always replace any blown fuse with the same rated fuse.**

The DC fuse should only trip if there is a short circuit in the supply itself (downstream short circuits or overloads will trip individual Polyswitches). This could occur if the F1-H terminal block somehow contacts DC negative. Alternately, if you are not using the "P" terminals for downstream wiring but are using the "H" terminal to operate an individual, high current, downstream load, a short circuit or overload could trip the DC fuse.

Securitron's power supply family contains an additional safety feature which is **automatic shut off** in the event of a DC short circuit or overload. This is often called a "crowbar" circuit. When you are using the power supply **without batteries**, a DC short circuit will usually cause the power supply to shut itself off rather than tripping any fuse or Polyswitch. If batteries are implemented, however, they will attempt to drive into the load as soon as the short circuit or overload occurs and the fuses and/or Polyswitches will trip to maintain safety.

If this happens there is a **reset procedure. First, correct the overload condition. Next, remove all current from the Polyswitch for a period of 10 seconds**. You do this by moving the associated slide switch to the "off" position. Then wait 10 seconds, return the slide switch to "on" and operation will return to normal. If you haven't corrected the overload, naturally the Polyswitch will trip again but you must always de-power and re-power the Polyswitch to reset it.

The multiple safety features of these power supplies can be confusing so the following chart provides summary information on operation of the safety features when the power supply is used with and without batteries.

	AC FUSE	DC FUSE	POLYSWITCH	AUTO SHUTOFF
WITHOUT BATTERIES	Will trip only if internal transformer shorts (all LED's will be out)	Generally will not trip. Supply will go into auto shut-off in case of overload (all LED's will be out)	Generally will not trip. Supply will go into auto shut-off in case of overload* (all LED's will be out).	Will generally occur in the case of any short or overload (all LED's will be out)
WITH BATTERIES	Will trip only if internal transformer shorts (all LED's will be out)	Will trip if terminal F1, F2 or H shorts to negative or in case of overload when terminal H is used as single output (all LED's will be out)	Will trip in the event of individual zone short or overload (individual zone LED will be out or dim)	Batteries drive into short or overload which trips another safety feature unless overload current is less than fuse or Polyswitch rating.

* a Polyswitch can individually trip in an overload condition without batteries in the special case where the overload current is greater than the Polyswitch trip current (2.5 Amps) but less than the power supply output capacity. This is unusual. A pure short circuit is more common and this will put the supply into Auto shut-off.

LED DIAGNOSTIC TABLE		
RED (DC) on	GREEN (AC) on	
FRONT PANEL	CONTROL BOARD	DESCRIPTION
ON	ON	Normal Operation
ON	OFF	Battery backup is powering output
OFF	ON	No DC output
OFF	OFF	No AC, no battery or discharged battery

VOLTAGE SUPERVISION DIAGNOSTIC TABLE		
AC FAIL TERMINALS	LOW BATT TERMINALS	DESCRIPTION
Continuity between C and NO	Continuity between C and NO	Normal Operation
Continuity between C and NO	Continuity between C and NC	AC present, no DC ouput
Continuity between C and NC	Continuity between C and NO	No AC, battery is powering unit
Continuity between C and NC	Continuity between C and NC	No AC, no battery or discharged battery

FIG. 1: POWER SUPPLY WIRING WITH CCS8M BOARD

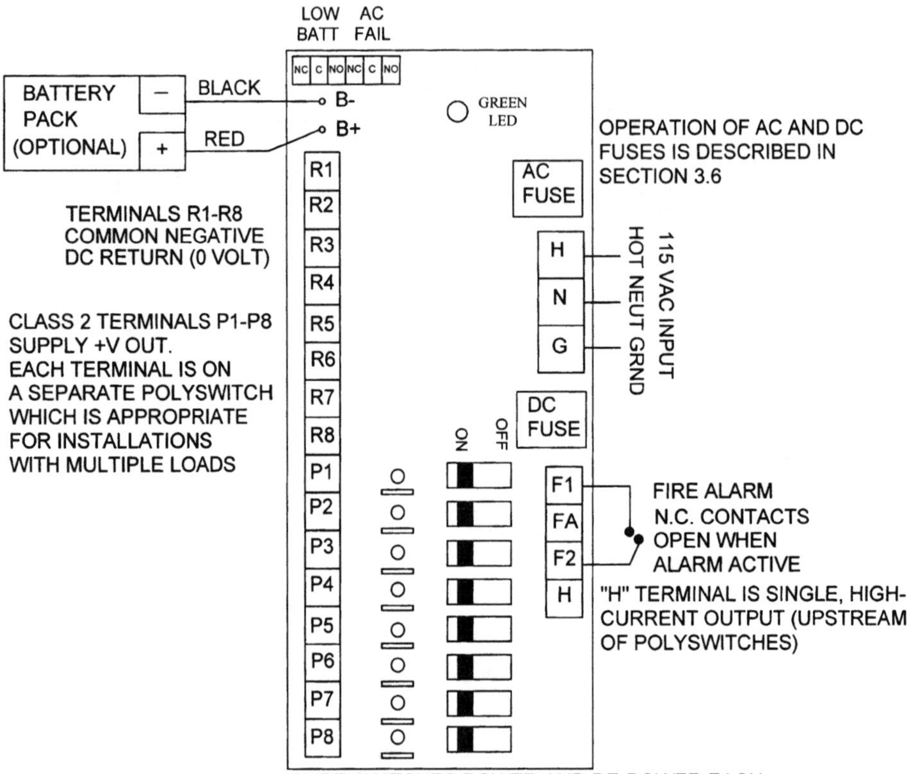
LOW BATT
AC FAIL
NC C NO NC C NO
BATTERY PACK (OPTIONAL)
–
+
BLACK
RED
B-
B+
GREEN LED
OPERATION OF AC AND DC FUSES IS DESCRIBED IN SECTION 3.6
AC FUSE
TERMINALS R1-R8 COMMON NEGATIVE DC RETURN (0 VOLT)
R1
R2
R3
R4
R5
R6
R7
R8
H
N
G
115 VAC INPUT
HOT NEUT GRND
CLASS 2 TERMINALS P1-P8 SUPPLY +V OUT.
EACH TERMINAL IS ON A SEPARATE POLYSWITCH WHICH IS APPROPRIATE FOR INSTALLATIONS WITH MULTIPLE LOADS
DC FUSE
ON
OFF
P1
P2
P3
P4
P5
P6
P7
P8
F1
FA
F2
H
FIRE ALARM N.C. CONTACTS OPEN WHEN ALARM ACTIVE
"H" TERMINAL IS SINGLE, HIGH-CURRENT OUTPUT (UPSTREAM OF POLYSWITCHES)

SLIDE SWITCHES POWER AND DE-POWER EACH "P" OUTPUT. LED'S SHOW OUTPUT STATUS

3.7 BATTERY CHARGING CAPABILITY

A resistor and diode are present on the CCS8M board which, together with the power supply, constitute a battery charging circuit appropriate for standby rated sealed lead acid or gel cell batteries. **Dry cell or NICAD batteries must not be used.** Batteries are an option. The power supply can be used with or without them. The battery pack of the appropriate voltage is connected to the red and black flying leads following correct polarity. In the event of a line voltage power failure, the batteries will automatically drive the load at the same DC voltage. If the emergency release terminals are opened, battery power will, however, be blocked just as normal power from the power supply would be.

The components utilized on the CCS8M board for battery charging function for battery packs up to **20 amp hours in capacity** whether 12 or 24 volts. Larger battery packs can be handled but Securitron must be informed so that the board may be modified. Consult (Figure 2) to calculate the correct battery pack based on desired backup time and the current drawn by the load. **For proper battery charging, the power supply must be set at 27 volts for a 24 volt system, and 13.5 volts for a 12 volt system.** Securitron power supplies are factory set to this level and if it is not maintained, the batteries will not hold their full capacity, and may even be damaged.

FIG 2: CHART TO DETERMINE SIZE OF BATTERY PACK

BACKUP TIME DESIRED

TOTAL CURRENT DRAWN	MIN	1 HR	2 HR	4 HR	UL STD.	8 HR	16 HR	24 HR	48 HR	72 HR
150 MA	4 AH	4 AH	4 AH	4 AH	4 AH	4 AH	4 AH	8 AH	8 AH	12 AH
300 MA	4 AH	4 AH	4 AH	4 AH	4 AH	4 AH	8 AH	12 AH	16 AH	24 AH
500 MA	4 AH	4 AH	4 AH	4 AH	4 AH	8 AH	12 AH	16 AH	24 AH	36 AH
1 A	4 AH	4 AH	4 AH	8 AH	12 AH	12 AH	20 AH	24 AH	48 AH	72 AH
2 A	4 AH	4 AH	8 AH	12 AH	20 AH	20 AH	36 AH	48 AH	100AH	150AH
3 A	4 AH	8 AH	12 AH	16 AH	24 AH	28 AH	52 AH	72 AH	150AH	240AH
4 A	4 AH	8 AH	16 AH	20 AH	32 AH	36 AH	72 AH	100AH	200AH	300AH
5 A	4 AH	12 AH	16 AH	24 AH	40 AH	44 AH	84 AH	120AH	240AH	360AH
7.5 A	4 AH	16 AH	20 AH	36 AH	60 AH	72 AH	130AH	180AH	360AH	480AH
10 A	4 AH	20 AH	28 AH	48 AH	72 AH	100AH	180AH	240AH	480AH	720AH

"MIN" TIME REFERS TO FACILITY USING A GENERATOR WHERE THE BATTERIES ARE ONLY REQUIRED TO OPERATE THE SYSTEM FOR A FEW MINUTES

U.L. STANDARD REQUIRES 4 HOURS OF BATTERY OPERATION FOLLOWED BY A 24 HOUR RECHARGE PERIOD AND THEN A SECOND 4 HOURS OF OPERATION

STANDARD SECURITRON POWER SUPPLIES CAN ONLY CHARGE UP TO A 20AH PACK. IF A LARGER PACK IS CALLED FOR, THE FACTORY MUST BE ALLERTED TO SUPPLY MODIFIED EQUIPMENT. LARGER PACKS ARE SHOWN IN ITALICS IN THE CHART.

BATTERIES MUST BE SEALED LEAD ACID OR GEL CELL TYPES. DRY CELLS WILL NOT RECHARGE AND WILL BE DAMAGED.

THIS CHART IS ONLY VALID IF BATTERIES ARE OPERATED AT ROOM TEMPERATURE. IN A COLD ENVIRONMENT, CAPACITY IS REDUCED.

BATTERIES SHOULD BE REPLACED AFTER 5 YEARS OF USE.

3.8 CODE APPROVED WIRING METHODS

Note that these units are **Class 2 rated**. This means that the individual DC outputs of the supplies (on "P" terminals) are current limited and can pose neither a high voltage nor high energy hazard outside of the enclosure. Electrical building codes in most jurisdictions permit Class 2 wiring to be done "in the open" rather than in conduit. To **maintain the Class 2 ratings** on the "P" terminal outputs, **never connect them together** to obtain higher capacity. If you require higher capacity use the "H" terminal but the "H" terminal is high current (not Class 2) and must be in conduit. The line voltage wiring coming into the unit also must be in conduit as it poses a high voltage hazard. **Be sure to check with your local building department** to make sure you are complying with applicable wiring codes before installing these units.

INSTALLATION AND OPERATING INSTRUCTIONS
For Model M38 and M68 Magnalocks
(Versions "L", "S" and "LS" with "D" and "T" Options)

1. INTRODUCTION

Securitron's Model M38 and M68 Magnalocks are fail safe electromagnetic locks designed for *interior use in areas which require controlled access or egress. Each unit includes an interlocking mounting bracket and a power and control wiring access compartment which provides the added convenience of wiring "at-the-lock". The sleek, low-profile design provides a professional looking, unobtrusive integration into a wide range of architectural applications and can be easily retrofitted to replace existing magnetic lock systems.

M38 and M68 Magnalocks are not designed, recommended or approved for exterior (outdoor) use. Interior use is described as any door installation to the interior of a building (i.e. offices, work or storage rooms) or to the interior side of any perimeter exterior door.

2. SPECIFICATIONS

M38:
Holding Force: 600 Lbs. [272 Kg]
Dimensions:
 Inches: 10-1/2"L X 2-9/16"H X 1-9/16"D
 Millimeters: 267 L X 65 H X 40 D
Electrical:
 12 Volts DC:
  Current Requirement: 300 – 325 mA
  Power Consumption: 3.6 Watts
 24 Volts DC:
  Current Requirement: 150 – 175 mA
  Power Consumption: 3.6 Watts

M68:
Holding Force: 1,200 Lbs. [544 Kg]
Dimensions:
 Inches: 10-1/2"L X 3-5/8"H X 1-13/16"D
 Millimeters: 267 L X 92 H X 46 D
Electrical:
 12 Volts DC:
  Current Requirement: 250 – 275 mA
  Power Consumption: 3.0 Watts
 24 Volts DC:
  Current Requirement: 125 -150 mA
  Power Consumption: 3.0 Watts

3. PRODUCT OVERVIEW

Upon unpacking this product, an inventory should be made to ensure that all the required components and hardware have been included. Along with these instructions and the installation template, the lock assembly (M38 Shown) should include the following items:

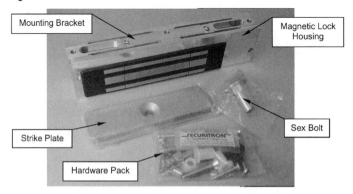

© Copyright, 2004, all rights reserved • Securitron Magnalock Corp., 550 Vista Blvd., Sparks NV 89434, USA
Tel: (775) 355-5625 • (800) MAGLOCK • Fax: (775) 355-5636 • Website: www.securitron.com

An ASSA ABLOY Group company |ASSA ABLOY

4. RECOMMENDED TOOLS

Hammer	Wrench, 1/2" box-end
Center punch	Pliers, vise grip
Drill motor	Screwdrivers: #1 and #3 Phillips
Drill bits: 3/16", 7/32" 3/8" and 1/2"	Hex (Allen) wrenches: 1/16" (provided) & 3/16"

5. INSTALLATION INSTRUCTIONS

5.1. Pre-Installation Survey

Due to the variety of mounting configurations available with this product, it is strongly recommended that an initial survey and assessment be made of the physical area to which the Magnalock will be installed. A determination of the optimal method of mounting should be made prior to installation with considerations made to the following:

A. Physical strength of mounting area. The structural integrity of mounting surfaces must be strong enough to meet or exceed the holding force of the Magnalock.

B. Protection of the lock from external attack. The Magnalock and the wiring must be protected to a reasonable degree from potential damage due to intruders or vandals.

C. Convenience and accessibility of area to be protected. The Magnalock assembly should be installed in a location that will not hinder or create a potential safety hazard to authorized personnel routinely accessing the protected area.

The mounting configuration addressed in detail in this manual will be for an outswinging door. All hardware required for the outswinging mount configuration has been included with this unit. For the inswinging version of mounting, a bracket accessory Top Jam Kit (TJ-38 or TJ-68 Series) is required. Both kits are available through Securitron. The following illustrations show two basic mounting configurations:

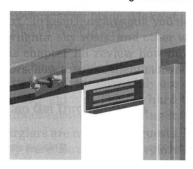
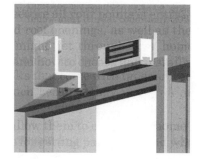

Outswinging Door Mount   Inswinging Door Mount (with TJ-38 Kit)

The Magnalock should normally be mounted under the door frame header in the corner furthest from the hinges. Most commonly the lock is positioned horizontally but vertical orientation may also be considered. In some cases for example, the horizontal header on an aluminum frame glass door is not as strong as the vertical extrusion, so vertical mounting would obviously be preferred. This (outswinging) type of installation places the Magnalock on the opposite side of the door as the door swings away from the lock. This configuration should be used for all facility exit doors (otherwise the lock would be located on the outside of the building). For interior doors, it is also recommended that the lock be mounted in this manner unless security planning anticipates a physical assault on the lock from that side of the door in which case the inswinging mounting kit (TJ-38 or TJ-68) should be obtained from Securitron.

5.2. Door Lock Installation

Figure 1 below is a typical cross section view of an M38 installed to a metal door frame in the outswinging door mount configuration:

Note: Reference dimensions are in parenthesis – see the product mounting template for clarification.

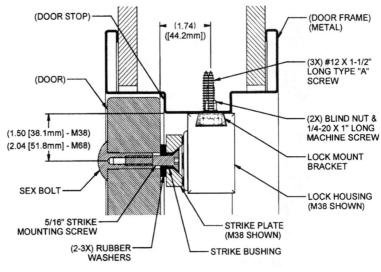

Figure 1

5.2.1. Door Preparation

Figure 2 below is an exploded illustration of the assembly of the strike to the door:

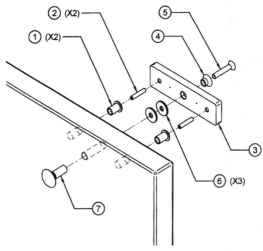

Figure 2

ITEM	QUANTITY	PART NUMBER	NOMENCLATURE/DESCRIPTION
1	2	560-12050	Strike Roll Pin Bushing, Clear, Nylon
2	2	330-10800	Roll Pin, 1/4" X 1-1/4" Long, Steel (Z/P)
3	1	380-30300 775-60400	Strike Plate, M32/M38 (shown) Strike Plate, M62/M68
4	1	330-12000	Strike Mounting Bushing, White, Delrin
5	1	300-13600	Socket Flat Head Screw, 5/16-18 UNC X 1-3/4" Long, Steel (Z/P)
6	3	310-11100	Washer, 1" O.D. X 11/32" I.D. X 1/8" Thick, Neoprene
7	1	330-12600	Sex Bolt, 5/16-18 UNC X 1-1/2" Deep, Steel (Z/P)

5.2.1.1.   Strike Installation

Using **Figure 2**, the template provided and the following step-by-step installation instructions, complete the installation of the strike into the door.

1. Fold the template and place into position on door/frame at desired mounting location. Mark the strike center hole on the door and the center position of the two (2) slotted mounting bracket holes on the door frame.
2. Drill one (1) 3/8" [9.5mm] diameter hole through door (at mark for strike center hole), then drill 1/2" [12.7mm] diameter through door from opposite side for sex bolt (**Item #7**).
3. From the side of the door that the strike will be mounted, drill two (2) 1/2" [12.7mm] diameter X 1-3/8" [34.9mm] deep holes for the roll pin bushings (**Item #1**).
4. Insert the two (2) strike roll pin bushings into the holes in the door (either side of strike center hole).
5. Obtain the strike (**Item #3**) and place on a clean, flat surface. Using a hammer, insert and **lightly** tap to install the two (2) roll pins (**Item #2**) into the two (2) 1/4" [6.4mm] diameter recesses of the strike. **Note:** Care should be taken when installing roll pins into the strike. Excessive impact to the strike or over-driving the pins into the recesses may distort the contact surface of the strike which will affect the holding force of the lock.
6. Insert the strike mounting bushing (**Item #4**) and the 5/16-18 UNC X 1-3/4" long socket flat head screw (**Item #5**) through the center hole of the strike.
7. Install two (2) of the neoprene washers (**Item #6**) over the screw on the opposite side (roll pin side) of the strike. **Note:** A third washer is provided but is normally not used. It may be used however, if required to take up any additional spacing that cannot be taken up by adjusting the lock mount bracket.
8. Insert the strike mounting screw (with washers) into the 1/2" [12.7mm] diameter strike center hole in the door while aligning the roll pins of the strike assembly into the roll pin bushings.
9. Start the sex bolt (**Item #7**) into the 1/2" [12.7mm] diameter hole from the opposite side of the door. **Do not** fully seat the sex bolt against the surface of the door at this time, but thread the sex bolt onto the end of the strike mounting screw. Using a 3/16" hex wrench, continue to thread the strike mounting screw into the sex bolt while ensuring that the assembly maintains a straight (perpendicular) alignment to the door.
10. Using a hammer and a 3/16" hex wrench, gradually tighten the screw into the sex bolt until it is snug, and then tap the head of the sex bolt toward the face of the door. Keep repeating this procedure to slowly "walk" the sex bolt into place (head against door face) – this ensures a straight (perpendicular) alignment of the strike assembly to the door. **Note: Do not over tighten.** When the strike is fully installed there should be some play or flexing of the strike around the washers. This allows the lock to pull the strike into the correct alignment for maximum holding force.

5.2.2. Door Frame Preparation

Figure 3 below is an exploded illustration of the assembly of the lock to a metal door frame:

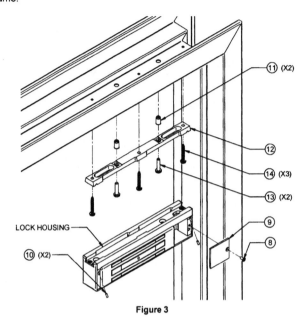

Figure 3

Figure 4 below is an exploded illustration of the assembly of the lock to a wood door frame:

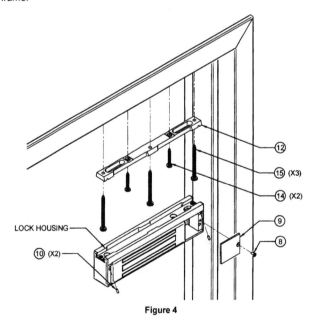

Figure 4

ITEM	QUANTITY	PART NUMBER	NOMENCLATURE/DESCRIPTION
8	1	(Included)	Phillips Flat Head Screw, 8-32 UNC
9	1	(Included)	Lock Housing Front Cover
10	2	(Included)	Socket Set Screw, 6-32 UNC
11	2	320-10800	Blind Nut, 1/4-20 UNC, 1-Piece
12	1	(Included)	Standard Mounting Bracket
13	2	300-12403	Phillips Pan Head Screw, 1/4-20 UNC X 1" Long, Steel (Z/P)
14	3	300-13010	Phillips Pan Head Screw, #12 X 1-1/2" Long, Type "A", Steel (Z/P)
15	3	300-13110	Phillips Pan Head Screw, #14 X 3" Long, Type "A", Steel (Z/P)

5.2.2.1.   Lock Installation

Using **Figure 3** (for metal door frames) or **Figure 4** (for wood door frames) above, the template provided and the following step-by-step installation instructions, complete the installation of the lock onto the door frame.

1. Locate the two (2) marks applied to the door frame for the mounting bracket using the template in the previous strike installation section (Section 5.2.1.1, Step 1), then:
    a. **For Metal Door Frame:** Drill two (2) 3/8" [9.5mm] diameter holes through the door frame at these locations.
        i. Install blind nuts (**Item #11**) into the 3/8" [9.5mm] diameter holes in the door frame as follows:
            1. Using blind nut collapsing tool provided (with blind nut in place), insert the end of the blind nut into the hole.
            2. Using a hammer (if necessary) tap the nut in until the upper lip seats against the door frame surface.
            3. While holding the hex portion of the collapsing tool with a 1/2" box-end wrench or vise grip pliers, turn the socket head cap screw of the tool using a 3/16" hex wrench. **Note:** Maintain firm pressure toward the door frame mounting surface while collapsing the nut.
            4. Once the blind nut is adequately collapsed, remove the tool from the nut by backing the cap screw out of the blind nut thread.
            5. Install the second blind nut onto the collapsing tool (cap screw) and repeat sub-steps i1 through i4.
    b. **For Wood Door Frame:** Drill two (2) 3/16" [4.8mm] diameter holes X 1-1/4" [31.8mm] deep (minimum) into the door frame at these locations.
2. Remove the mounting bracket (**Item #12**) from the top (dovetail) area of the lock housing as follows:
    a. Using a #1 Phillips screwdriver, remove the 8-32 UNC flat head screw (**Item #8**) holding the lock housing front cover (**Item #9**) in place. Remove cover and set screw and cover aside for later assembly.
    b. Using a 1/16" hex wrench, back the two (2) 6-32 UNC socket set screws (**Item #10**) out far enough to allow free (sliding) movement of the mounting bracket. **Note:** It is not necessary to completely remove the set screws to allow removal of the bracket – **five (5) 360-degree counterclockwise turns** is normally far enough.
    c. Slide the bracket far enough to disengage the bracket from the top of the lock housing. **Note:** The bracket/lock interface can be disengaged by moving the lock (or bracket) 1" to 2" [25-50mm] laterally to one side or the other, then withdrawing away vertically – you do not have to slide the lock or bracket the full length of the housing to remove or install.

3. Place the mounting bracket onto the frame aligning the two (2) slotted holes in the bracket with the previously installed blind nuts (for metal frame) or drilled holes (for wood frame). Using a #3 Phillips screwdriver, lightly secure the mounting bracket to the frame with:
   a. **For Metal Door Frame:** Two (2) 1/4"-20 UNC X 1" long Phillips pan head screws (**Item #13**) provided.
   b. **For Wood Door Frame:** Two (2) #12 X 1-1/2" long Phillips pan head type "A" screws (**Item #14**) provided.
   Note: Do not fully secure (tighten) the screws at this point. Tighten screws to "snug" and back out as necessary to provide slight movement of bracket for adjustment.
4. Assemble the lock housing to the bracket by inserting and engaging the top (dovetail) feature of the lock with the mounting bracket. Slide the lock into proper position (centered on the mounting bracket).
5. Adjust the lock/bracket assembly in the slots of the bracket as necessary to establish proper relationship with (against) the strike.
6. Carefully remove the lock housing from the mounting bracket while attempting not to alter the position of the bracket on the door frame.
7. Using a #3 Phillips screwdriver, fully secure the two (2) screws (**Item #13 or #14**) installed in Step 3 above.
8. Using the installed mounting bracket as a guide, locate and drill three (3) holes into the door frame corresponding to the remaining three (3) open holes in the bracket:
   a. **For Metal Door Frame:** Drill three (3) 3/16" [4.8mm] diameter holes (through).
   b. **For Wood Door Frame:** Drill three (3) 7/32" [5.6mm] diameter holes X 2-3/4" [69.9mm] deep (minimum).
9. Using a #3 Phillips screwdriver, install the three (3) #12 screws (**Item #14** for metal frame) **or** #14 screws (**Item #15** for wood frame) through the bracket and into the holes in the door frame. Tighten the screws to secure the mounting bracket to the door frame.
10. Determine optimal location then drill a 3/8" [9.5mm] diameter hole through the door frame for the lock power cable. **Note:** This hole must align within the large slotted "window" in the top of the lock housing when the lock is installed on the mounting bracket.
11. After all electrical wiring has been completed, install the housing front cover (**Item #9**) and secure in place using the 8-32 UNC flat head screw (**Item #8**).

5.3. Lock with "D" (Door Position) Option - Installation

For locks provided with the "D" (Door Position) option, the magnetic (reed) switch is factory installed inside the control wiring access compartment cover.

When a lock has been previously installed and it has been later determined to upgrade to the "D" option in the field, the template provided in the upgrade kit illustrates proper location and installation of the switch to the inside of the compartment access cover.

Using **Figure 5** below, the template provided and the following step-by-step installation instructions complete the installation of the actuator block to the door adjacent to the strike plate.

1. After marking the mounting holes for the actuator block on the door using the mounting template provided with the lock (or upgrade kit), center punch these locations. The actuator block has hardware included for mounting in two types of configurations. Decide which of the following choices will work best for the application and then:

a. **If Mounting Using #8 Flat Head Screws (Recommended for Solid Core (Wood), Aluminum Frame or Hollow Metal Doors):** Drill two (2) 9/64" [3.6mm] diameter holes X 3/4" [19.1mm] deep (maximum) into the door at the center punched locations or;

b. **If Mounting Using #6 Pan Head Screws (Recommended for Aluminum Frame, Hollow Metal or the Top Jam Z-Bracket):** Drill two (2) 1/8" [3.2mm] diameter holes X 3/8" [9.5mm] deep (maximum) into the door at these locations.

2. Using a #1 Phillips screwdriver and the screws provided secure the actuator block to the door making sure that the **arrow of the label** on the backside of the block is **facing toward the mounting edge of the lock**.

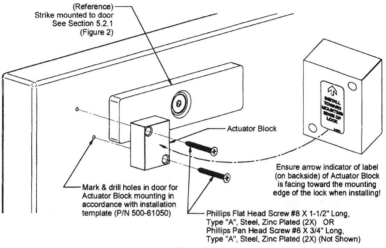

(Reference) Strike mounted to door See Section 5.2.1 (Figure 2)

Actuator Block

Ensure arrow indicator of label (on backside) of Actuator Block is facing toward the mounting edge of the lock when installing!

Mark & drill holes in door for Actuator Block mounting in accordance with installation template (P/N 500-61050)

Phillips Flat Head Screw #8 X 1-1/2" Long, Type "A", Steel, Zinc Plated (2X)   OR Phillips Pan Head Screw #6 X 3/4" Long, Type "A", Steel, Zinc Plated (2X) (Not Shown)

Figure 5

5.4. Inswinging Door Lock Installation

As previously mentioned, additional hardware and brackets are required for mounting the lock system in the inswinging door configuration. The brackets, hardware and instructions required for the installation of the lock in this arrangement are available in a Top Jam Kit (TJ-38 or TJ-68 Series) which may be obtained through Securitron Magnalock Corporation or one of our authorized representatives.

6. OPERATIONAL INSTRUCTIONS

The M38/68 series Magnalocks are direct holding fail safe electromagnetic locks. Both incorporate Securitron's unique dual voltage system. Simply apply 12 or 24 volts DC observing polarity (See **Figure 6**) and the locks will energize to their respective holding forces of 600 lbs. [272 Kg] for the M38 and 1200 lbs. [544 Kg] for the M68.

The **"L"** option adds the ability to visually inspect the lock and determine via a red LED that the lock is in fact energized.

The addition of the **"S"** option provides a lock status sensing signal which indicates that the lock is secure and gives you the added ability of a SPDT dry contact that you can interface to either the access control system or intrusion detection system (See **Figure 7**). The **"S"** option will only report the door is secure if two criteria are met. One - the lock must be powered and two - the strike armature must be in full unobstructed contact with the lock face.

Add the two options together **"LS"** and get both the dry contact output and a visual reference via a bi-color LED (red = door powered and locked, green = door powered but not locked and off = door not powered or locked).

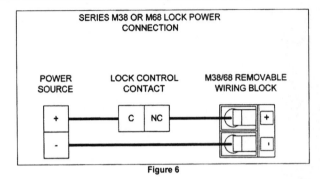

Figure 6

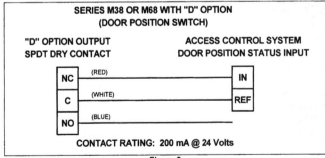

Figure 7

The addition of the **"D"** option incorporates a door position sensing switch which indicates when the door is outside of the sensing range of the lock (3/8"-3/4") and provides the added ability of a SPDT dry contact that can be interfaced with either an access control system or intrusion detection system (See **Figure 8**). The **"D"** option will report the door position whether the lock is powered (secured) or not.

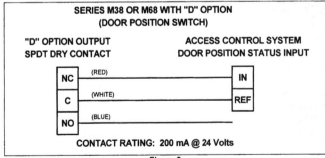

Figure 8

The **"T"** suffix option implements a small micro switch to create a tamper resistant enclosure of the control wiring access compartment. At approximately .03" movement of the access cover away from the lock housing (usually less than a full turn of the mounting screw) a signal is transmitted to indicate that access is being attempted to the lock control wiring compartment. (See **Figure 9** for wiring diagram).

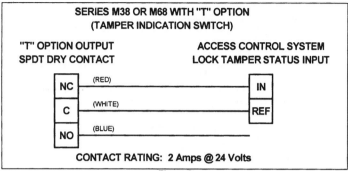

Figure 9

The **"OS"** suffix option available for M68 series locks provides a modified "Offset" strike to shift the strike contact face 1/4" [6.4mm] further away from the door frame stop than the standard strike. This can be helpful in the installation to aluminum frame glass doors where the height of the aluminum rail at the top of the door is not sufficiently wide enough to accommodate the standard sex bolt/strike installation. It should be noted that an approximate 10% loss in holding force results from the use of this strike because of its offset position however, this should not be significant in that aluminum frame glass doors are not high security barriers. The offset strike is included with locks ordered with the **"OS"** suffix or may be ordered separately.

A "Split Strike" configuration option is available for M68 series locks ordered with the **"SS"** suffix. This allows one lock to be used to secure two doors in a double door configuration. In this method, a single Magnalock is mounted near the center of the header and a half size strike is provided for mounting to each door/leaf. This arrangement will also reduce the holding force to about 550 Lbs. (250Kg) per door/leaf. Locks ordered with the **"SS"** suffix include the two (split) strikes or they may be supplied separately.

7. MAINTENANCE

To keep the Magnalock in top working order, especially a Magnalock equipped with the **"S"** (Senstat) option, we recommend taking a clean cloth and rubbing alcohol or a non-abrasive cleaner and wiping down the face of the Magnalock and the strike once every six months.

8. APPENDICES

A. WIRE GAUGE SIZING

If the power supply is distant from the lock, voltage will be lost (dropped) in the connecting wires so that the Magnalock will not receive full voltage. The following chart shows the minimum wire gauge that will hold voltage drop to an acceptable 5% for different lock to power supply distances. Proper use of the chart assumes a dedicated pair of wires to power each Magnalock (no common negative). Note that a Magnalock operating on 24 volts is a much better choice for long wire runs as it has 4 times the resistance of a 12 volt installation. Also note that the correct calculation of wire sizing is a very important issue as the installer is responsible to insure that adequate voltage is supplied to any load. In multiple device installations, the calculation can become quite complex so refer to the following section *"Calculating Wire Gauge Sizing"* for a more complete discussion.

Distance	Gauge 12V	Gauge 24V	Distance	Gauge 12V	Gauge 24V
100 FT	24 GA	24 GA	800 FT	18 GA	22 GA
200 FT	22 GA	24 GA	1500 FT	14 GA	18 GA
400 FT	20 GA	22 GA	2000 FT	14 GA	18 GA

CALCULATING WIRE GAUGE SIZING

The general practice of wire sizing in a DC circuit is to avoid causing voltage drops in connecting wires that reduce the voltage available to operate the device. As Magnalocks are very low power devices, they can be operated long distances from their power source. For any job that includes long wire runs, the installer must be able to calculate the correct gauge of wire to avoid excessive voltage drops.

This is done by taking the current draw of the lock and multiplying by the resistance of the wire I x R = Voltage drop (i.e. 0.100A x 10.1 Ohms = 1.01 Volts dropped across the wire). For all intents and purposes it can be said that a 5% drop in voltage is acceptable so if this were a 24 Volt system (24 Volts x .05 = 1.2 Volts) a 1.01 Volt drop would be within tolerance.

To calculate the wire resistance, you need to know the distance from the power supply to the Magnalock and the gauge (thickness) of the wire. The following chart shows wire resistance per 1000 ft (305 meters):

Wire Gauge	Resistance/1,000 ft	Wire Gauge	Resistance/1,000 ft
8 Gauge	.6 Ohms	16 Gauge	4.1 Ohms
10 Gauge	1.0 Ohms	18 Gauge	6.4 Ohms
12 Gauge	1.6 Ohms	20 Gauge	10.1 Ohms
14 Gauge	2.5 Ohms	22 Gauge	16.0 Ohms

B. CONSIDERATIONS FOR MAXIMUM PHYSICAL SECURITY

M38 and M68 Magnalocks carry rated holding forces of 600 lbs. [272 Kg] and 1,200 lbs. [544 Kg] respectively. There are several installation and application variables to be considered which affect the security level that may be obtained while using a Magnalock.

In the case of wooden doors (other than solid hardwoods), aluminum frame glass doors, and hollow metal doors, the M38 should be employed in a "traffic control" mode. This is because a determined assault on these types of door/lock configurations has the potential of "popping" this model open. The M68 is generally stronger that the door itself. Users have logged periodic cases of an assault where the door has been destroyed leaving the Magnalock intact and still retaining the piece of the door where the strike was mounted.

A pry bar may be used to try to pry the door open. However, what generally occurs is that the door will experience material failure. The pry bar tears (in the case of wood) or bends (metal) the door material without defeating the lock. The general fact that a Magnalock mounts on the other side of the door from the attacker is an important contributor to its strong resistance to assault.

The concept of preferring a door that gives also affects the issue of physical security on different door types. Oddly enough, the characteristics that make an inswinging door strong can make it more difficult to secure if it is stiff or rigid. Steel doors and most particularly solid steel doors such as may be found in prisons, transmit impact much more effectively to the lock and as such may be defeated by repeated leg blows or by charging the door. If the end user has such doors and/or a security environment where such determined attacks can be foreseen on the doors, it is his responsibility to ensure that the Magnalock's strength is adequate for the application. Selection of the model 68 is strongly recommended with another technique being the use of multiple locks.

These Magnalocks have been carefully designed to be installed with the included fasteners which substantially exceed the strength of the lock. If the installer substitutes for any of the factory included fasteners or hardware for any reason, the strength capability of the installation may be compromised. It should be noted that a key component to the successful installation of this product is the sex bolt. The head of the sex bolt is the only part of the lock assembly that is accessible to an attacker.

It is also vital when mounting the magnet (lock housing) into a metal frame to use the included machine screws and steel blind nuts. An alternative technique has been to use sheet metal screws which some installers feel is easier. This is extremely ill advised as the magnet receives a torquing force each time that the door is closed, which over time will work the sheet metal screws loose. It has been determined that sheet metal screws may be considered acceptable, although not preferred, if the frame header is made of steel. Indeed this is mandated when Securitron's concrete header bracket is utilized. On an aluminum header however, mounting with sheet metal screws is dangerous as the steel screw threads will eventually tear through the relatively soft aluminum.

To fully complete an installation that maximizes the effectiveness of the included fasteners, the thread locker compound (provided) should be used to prevent the threads from loosening over time.

C. TROUBLESHOOTING

PROBLEM: No magnetic attraction between magnet and strike plate.

First be sure the Magnalock is being correctly powered with DC voltage. This includes connecting the power wires with correct polarity. Positive must go to red and negative to black. If the magnet body is wired in reverse polarity, it will not be damaged, but it will not operate. If the unit continues to appear dead, it must be electrically checked with an Ammeter. It must be powered with the correct input voltage and checked to see if it draws the specified current. If the unit meters correctly, it is putting out the correct magnetic field and the problem must lie in the mounting of the strike.

PROBLEM: Reduced holding force.

This problem usually expresses itself in terms of being able to kick the door open or to open it with a shoulder. Check the strike and magnet face to see if some small obstruction is interfering with a flat fit. Even a small air gap can greatly reduce the holding force. Another possibility is if the strike plate has a dent on it from being dropped for example. Remove the strike from the door and try to rock it on the magnet face to insure that it is flat. If the strike and magnet are flat and clean, the cause is nearly always improper mounting of the strike in that the strike is mounted too rigidly. The strike **must** be allowed to float around the rubber washer stack which must be on the strike center mounting screw. The magnet then pulls it into flat alignment. To correct the problem, try loosening the strike mounting screw to see if the lock then holds properly. Another possibility is if you are operating the lock on AC instead of DC or on half wave rectified DC (transformer + single diode). Half wave rectified DC is unacceptable; you must at a minimum employ full wave rectified DC (transformer + bridge).

PROBLEM: The Senstat output does not report secure.

Because of the simplicity of Securitron's patented Senstat design, this is almost always a case of the lock status sensor doing its job. It is not reporting secure because a small obstruction or a too stiffly mounted strike is causing the Magnalock to hold at a reduced force. The problem is corrected by cleaning the surfaces of the magnet and strike (see Maintenance Section 7) or establishing proper play in the strike mounting. If this does not work, you can verify function of the Senstat feature as follows: Note that there are two (2) thin vertical lines on the magnet face that can be said to separate the core into three (3) sections from left to right. The Senstat output is created by the strike establishing electrical contact between the leftmost and rightmost core segments. With the lock powered, use a pair of scissors and press the points respectively into the leftmost and rightmost core segments. The Senstat output should then report secure. This shows that the problem lies in the strike not making correct flat contact with the magnet face. If the scissors technique does not cause the lock to report secure, check to see if there is a broken Senstat wire. If this is not the case, the lock must be returned to the factory for replacement.

PROBLEM: The lock does not release.

When power is removed from it, the Magnalock must release. Therefore the complaint of "lock will not release" is either mechanical bonding via vandalism or a failure to completely release power. By mechanical bonding, we simply mean that glue has been applied between the strike plate and magnet as a prank. Failure to completely release power is generally a wiring integrity problem. What happens is that an upstream switch removes power from the wires going to the Magnalock, but through an installation error, the wires have their insulation abraded between the switch and lock so that partial or full power can leak in from another Magnalock or other DC device with similarly abraded wiring. This is most likely to occur at the point where the wire cable leaves the Magnalock case and enters the door frame. Another area is via an improper splice on wiring in conduit. Either a metal door frame or the metal conduit is capable of leaking power between multiple devices with abraded wires, thereby bypassing switches. A good way to check this electrically (as opposed to visually removing and inspecting the wires) is to use a meter and check for leakage between the power supply positive or negative and the door frame and conduit. Magnalocks should be powered by isolated DC voltage without any earth ground reference to positive or negative.

PROBLEM: The lock rusts.

Both the magnet core and strike plate are plated and sealed following a military specification. If rusting appears, the most common cause is that improper cleaning (with steel wool for instance) has occurred and this has stripped off the relatively soft plating. Once the plating has been removed, it cannot be restored in the field, so the lock will have to be periodically cleaned and coated with oil or other rust inhibitor. A rusty Magnalock will still function but at reduced holding force. If the product is installed in a heavily corrosive atmosphere, such as near the ocean, it will eventually rust even with non-abrasive cleaning. The only answer then becomes continued periodic removal of the rust.

PROBLEM: Apparent electronic noise interference with the access control system.

Electric locks, being inductive devices, return voltage spikes on their power wires and also emit microwave radiation when switched. This can interfere with access control electronics causing malfunctions. Access control contractors often employ installation techniques designed to isolate the access control electronics from the electric lock. These include separate circuits for the lock, shielded wiring and other techniques. These techniques will vary with the sensitivity of the access control system electronics and should, of course be followed. Note that Magnalocks include internal electronics which suppress both inductive kickback and radiation. They have been extensively tested and accepted by numerous access control manufacturers and have been used in thousands of installations without incident. An apparent noise problem is therefore usually not caused by the Magnalock. The access control equipment may be itself faulty or have been installed improperly. One problem can arise with the Magnalock. If the Senstat version is being used, the strike plate (which passes current) must be isolated from a metal door and frame. Securitron supplies insulating hardware to accomplish this but the hardware may not have been used or the strike may be scraping against the header for instance. Check for full isolation between the strike and the door frame (when the door is secured) with an Ohmmeter. The presence of lock voltage potential in the door frame can interfere with the ground reference of access control system data communication and therefore cause a problem.

IF YOUR PROBLEM PERSISTS
CALL SECURITRON TOLL FREE
1-800-MAG-LOCK

9. WARRANTY

MAGNACARE® LIMITED LIFETIME WARRANTY

SECURITRON MAGNALOCK CORPORATION warrants that it will replace at customer's request, at any time for any reason, products manufactured and branded by SECURITRON.

SECURITRON will use its best efforts to ship a replacement product by next day air freight at no cost to the customer within 24 hours of SECURITRON's receipt of the product from customer. If the customer has an account with SECURITRON or a valid credit card, the customer may order an advance replacement product, whereby SECURITRON will charge the customer's account for the price of the product plus next day air freight, and will credit back to the customer the full amount of the charge, including outbound freight, upon SECURITRON's receipt of the original product from the customer.

SECURITRON's sole and exclusive liability, and customer's sole remedy, is limited to the replacement of the SECURITRON product when delivered to SECURITRON's facility (freight and insurance charges prepaid by customer). The replacement, at SECURITRON's sole option, may be the identical item or a newer unit which serves as a functional replacement. In the event that the product type has become obsolete in SECURITRON's product line, this MAGNACARE warranty will not apply. This MAGNACARE warranty also does not apply to custom, built to order, or non-catalog items, items made by others (such as batteries), returns for payment, distributor stock reductions, returns seeking replacement with anything other than the identical product, or products installed outside of the United States or Canada. This MAGNACARE warranty also does not apply to removal or installation costs.

SECURITRON will not be liable to the purchaser, the customer or anyone else for incidental or consequential damages arising from any defect in, or malfunction of, its products. SECURITRON does not assume any responsibility for damage or injury to person or property due to improper care, storage, handling, abuse, misuse, or an act of God.

EXCEPT AS STATED ABOVE, SECURITRON MAKES NO WARRANTIES, EITHER EXPRESS OR IMPLIED, AS TO ANY MATTER WHATSOEVER, INCLUDING WITHOUT LIMITATION THE CONDITION OF ITS PRODUCTS, THEIR MERCHANTABILITY OR FITNESS FOR ANY PARTICULAR PURPOSE.

Glossary

Many security terms have popular synonyms and variant spellings. The entries in this glossary are based on a survey of security trade journals, reference books, technical manuals, supply catalogs, and trade association glossaries. When a main entry has two or more commonly used synonyms or spellings, they're listed in order of popularity. When two or more are equally popular, those are listed in order of age (from oldest usage to newest). In cases where two or more common terms are spelled alike but have different meanings, each term is listed in a separate entry.

ac Alternating current. Electric current that reverses its direction of flow at regular intervals. For practical purposes, alternating current is the type of electricity that flows throughout a person's house and is accessed by the use of wall sockets.

access code The symbolic data, usually in the form of numbers or letters, that allow entry into an access-controlled area without prompting an alarm condition.

access control Procedures and devices designed to control or monitor entry into a protected area. Frequently, access control is used to refer to electronic and electromechanical devices that control or monitor access.

Ace lock A term sometimes used to refer to any tubular-key lock. The term is used more properly to refer to the Chicago Ace Lock, the first brand name for a tubular-key lock.

actuator A device, usually connected to a cylinder, that, when activated, causes a lock mechanism to operate.

adjustable mortise cylinder Any mortise cylinder whose length can be adjusted for a better fit in doors of varying thickness.

AFTE Association of Firearm and Toolmark Examiners.

AHC Architectural Hardware Consultant, as certified by the Door and Hardware Institute.

all-section key blank The key section that enters all keyways of a multiplex key system.

ALOA Associated Locksmiths of America.

Americans with Disabilities Act A congressional act passed to ensure the rights of persons with various types of disabilities. Part of the act requires that buildings provide easy access to the physically disabled.

ampere (or amp) A unit of electric current.

angularly bitted key A key that has cuts made into the blade at various degrees of rotation from the perpendicular.

annunciator A device, often used in an alarm system, that flashes lights, makes noises, or otherwise attracts attention.

ANSI American National Standards Institute.

antipassback   A feature in some electronic access-control systems designed to make it difficult for a person who has just gained entry into a controlled area to allow another person to also use the card to gain entry.

antipick latch   A spring latch fitted with a parallel bar that is depressed by the strike when the door is closed. When the bar is depressed, it prevents the latch from responding to external pressure from lock-picking tools.

architectural hardware   Hardware used in building construction, especially that used in connection with doors.

armored front   A plate covering the bolts or set screws holding a cylinder to its lock. These bolts normally are accessible when the door is ajar.

ASIS   American Society for Industrial Security.

associated change key   A change key that is related directly to particular master key(s) through the use of constant cuts.

associated master key   A master key that has particular change keys related directly to its combination through the use of constant cuts.

ASTM   American Society for Testing and Materials.

audit trail   A record of each entry and exit within an access-controlled area.

automatic deadbolt   A deadbolt designed to extend itself fully when the door is closed.

automatic flush bolt   A flush bolt designed to extend itself when both leaves of a pair of doors are in the closed position.

auxiliary code   A secondary or temporary access code.

auxiliary lock   Any lock installed in addition to the primary lockset.

back plate   A thin piece of metal, usually with a concave portion, used with machine screws to fasten certain types of cylinders to a door.

backset   The horizontal distance from the edge of a door to the center of an installed lock cylinder, keyhole, or knob hub. On locks with rabbeted fronts, it is measured from the upper step at the center of the lock face.

ball bearing   1. A metal ball used in the pin stack to accomplish some types of hotel or construction keying. 2. A ball, usually made of steel, used by some lock manufacturers as the bottom element in the pin stack in one or more pin chambers. 3. Any metal ball used as a tumbler's primary component.

ball locking   A method of locking a padlock shackle into its case using ball bearing(s) as the bolt(s).

barrel key   A key with a bit projecting from a hollow cylindrical shank. The hollow fits over a pin in a lock keyway and helps to keep the key aligned. The key is also known as a *hollow-post key* or *pipe key.*

bell crank (or bellcrank)   A flat metal plate attached to connecting rods within the walls of an automobile's doors. The bell crank converts movement from vertical to horizontal and vice versa.

bell key   A key whose cuts are in the form of wavy grooves milled into the flat sides of the key blade. The grooves usually run the entire length of the blade.

bezel   A threaded collar commonly used to secure certain cylinder or lock assemblies.

BHMA   Builders Hardware Manufacturers Association.

bible   That portion of the cylinder shell normally housing the pin chambers, especially those of key-in-knob cylinders or certain rim cylinders.

bicentric cylinder   A cylinder that has two independent plugs, usually with different keyways. Both plugs are operable from the same face of the cylinder. It is designed for use in extensive master-key systems.

bidirectional cylinder   A cylinder that can be operated in a clockwise and counter-clockwise direction with a single key.

binary cut key   A key whose combination only allows for two possibilities in each bitting position: cut/no cut.

binary-type cylinder or lock   A cylinder or lock whose combination only allows for two possibilities in each bitting position.

bit   1. The part of the key that serves as the blade, usually for use in a warded or lever-tumbler lock. 2. To cut a key.

bit key (or bit-key)   A key with a bit projecting from a solid cylindrical shank. The key is sometimes referred to as a *skeleton key*. A bit key is used to operate a bit-key lock.

bit-key lock   A lock operated by a bit key.

bitting   1. The number(s) representing the dimensions of the key cut(s). 2. The actual cut(s) or combination of a key.

bitting depth   The depth of a cut that is made into the blade of a key.

bitting list   A listing of all the key combinations used within a system. The combinations are usually arranged in order of the blind code, direct code, and/or key symbol.

bitting position   The location of a key cut.

blade   The portion of a key that contains the cuts and/or milling.

blank   See *Key blank*. An uncut key.

blind code   A designation, unrelated to the bitting, assigned to a particular key combination for future reference when additional keys or cylinders are needed.

block master key (BM)   The one-pin master key for all combinations listed as a block in the standard progression format.

bored lock   Any lock that is installed by cross-boring two holes in a door; one hole in the face of the door and the other through the edge of the door. Common bored locks include deadbolt locks and key-in-knob locks.

boring jig   A tool temporarily mounted to a door to act as a template and drill guide for installing hardware.

bottom of blade   The portion of the blade opposite the cut edge of a single-bitted key.

bottom pin   A tumbler, usually cylindrical (also may be conical, ball-shaped, or chisel-pointed), that makes contact with the key.

bow   The portion of the key that serves as a grip or handle.

bow stop   A type of stop located near the key bow.

box strike   A strike equipped to line the bolt cavity for both aesthetic and protective purposes.

break-away padlock or frangible padlock   A padlock with a shackle that's designed to be easily broken off. Break-away padlocks are used commonly on fire safety equipment and on other items that must be readily accessible.

broach   1. A tool used to cut the keyway into the cylinder plug. 2. To cut the keyway into a cylinder plug with a broach.

building master key   A master key that operates all or most master-keyed locks in a given building.

burglar-resistant   A broad term used to describe locks, doors, or windows designed to resist attack by prowlers and thieves for a limited time.

bypass key   The key that operates a key override cylinder.

cam   1. A flat actuator or locking bolt attached to the rear of a cylinder perpendicular to its plug and rotated by the key.

2. A lock or cylinder component that transfers the rotational motion of a key or cylinder plug to the bolt mechanism of a lock. 3. The bolt of a cam lock.

cam lock   A complete locking assembly in the form of a cylinder whose cam is the actual locking bolt.

cap   1. A spring cover for a single-pin chamber. 2. A part that can serve as a plug retainer and/or a holder for the tailpiece. 3. To install a cap.

capping block   A holding fixture for certain interchangeable cores that aids in the installation of the caps.

case   The housing or body of a lock.

case (of a cylinder)   See *Shell*.

case (of a lock)   The box that houses the lock-operating mechanism.

case ward   1. A ward attached directly to or projecting from a lock case. 2. A ward or obstruction integral to the case of a warded lock.

central processing unit (CPU)   Also called a *central processor;* the section in a digital computer that contains the logic and internal memory units.

chamber   Any cavity in a cylinder plug and/or shell that houses the tumblers.

changeable-bit key   A key that can be recombined by exchanging and/or rearranging portions of its bit or blade.

change key   A key that operates only one cylinder or one group of keyed-alike cylinders in a keying system.

circuit   A complete path through which electricity flows to perform work.

circuit breaker   A device designed to protect a circuit by automatically breaking (or opening) the circuit when current flow becomes excessive.

CK   1. Change key. 2. Control key.

clutch   The part of a profile cylinder that transfers rotational motion from the inside or outside element to a common cam or actuator.

CMC   Certified Management Consultant, as certified by the Institute of Management Consultants.

code   A designation assigned to a particular key combination for reference when additional keys or cylinders might be needed.

code key   A key cut to a specific code rather than duplicated from a pattern key. It may or may not conform to the lock manufacturer's specifications.

code original key   A code key that conforms to the lock manufacturer's specifications.

combinate   To set a combination in a lock, cylinder, or key.

combination   The group of numbers that represent the bitting of a key and/or the tumblers of a lock or cylinder.

combination lock   A lock that may or may not be operated with a key but can be operated by inserting a combination of numbers, letters, or other symbols by rotating a dial or by pushing buttons.

combination wafer   A type of disk tumbler used in certain binary-type disk-tumbler key-in-knob locks. Its presence requires that a cut be made in that position of the operating key(s).

compensate drivers   To select longer or shorter top pins depending on the length of the rest of the pin stack in order to achieve a uniform pin-stack height.

complementary keyway   Usually a disk-tumbler keyway used in master keying. It accepts keys of different sections whose blades contact different bearing surfaces of the tumblers.

complex circuit   A combination of series and parallel circuits.

composite keyway   A keyway that has been enlarged to accept more than one key section, often key sections of more than one manufacturer.

concealed-shell cylinder   A specially constructed (usually mortise) cylinder. Only the plug face is visible when the lock trim is in place.

conductor   Material, such as copper wire, used to direct current flow.

connecting bar   A flat bar attached to the rear of the plug in a rim lock to operate the locking-bar mechanism.

constant cut   Any bittings that are identical in corresponding positions from one key to another in a keying system. They usually serve to group these keys together within a given level of keying and/or link them with keys of other levels.

construction breakout key   A key used by some manufacturers to render all construction master keys permanently inoperative.

construction core   An interchangeable or removable core designed for use during the construction phase of a building. The cores normally are keyed alike, and on completion of construction, they are to be replaced by the permanent system's cores.

construction master key (CMK)   A key normally used by construction personnel for a temporary period during building construction. It may be rendered permanently inoperative without disassembling the cylinder.

construction master keyed   Of or pertaining to a cylinder that is operated temporarily by a construction master key.

control cut   Any bitting that operates the retaining device of an interchangeable or removable core.

control key   1. A key whose only purpose is to remove and/or install an interchangeable or removable core. 2. A bypass key used to operate and/or reset some combination type locks. 3. A key that allows disassembly of some removable cylinder locks.

control lug   The part of an interchangeable- or removable-core retaining device that locks the core into its housing.

control sleeve   The part of an interchangeable-core retaining device surrounding the plug.

controlled cross-keying   A condition in which two or more different keys of the same level of keying and under the same higher-level key(s) operate one cylinder by design. For example, XAAl can be operated by AA2 but not by AB1. This condition could severely limit the security of the cylinder and the maximum expansion of the system when more than a few of these different keys operate a cylinder or when more than a few differently cross-keyed cylinders per system are required.

convenience key   A key that has the same cuts on two sides of its blade and can operate the lock with either side of the blade facing up in the keyway. A convenience key has cuts on two sides for convenience only because only one side of the key manipulates the tumblers and other parts involved in locking and unlocking.

core   A complete unit, often with a figure-eight shape, that usually consists of the plug, shell, tumblers, springs, plug retainer, and spring cover(s). It is used primarily in removable- and interchangeable-core cylinders and locks.

corrugated key   A key with pressed longitudinal corrugations in its shank to correspond to a compatibly shaped keyway.

CJS   Certified Journeyman Safecracker, as certified by the National Safeman's Organization.

CMS   Certified Master Safecracker, as certified by the National Safeman's Organization.

CPL   Certified Professional Locksmith, as certified by Associated Locksmiths of America.

CPO   Certified Protection Officer, as certified by the International Foundation for Protection Officers.

CPP   Certified Protection Professional, as certified by American Society for Industrial Security.

crash bar   See *Panic bar.*

cross bore   A hole drilled into the face of a door where a bored or interconnected lockset is to be installed.

cross-keying   The deliberate process of combinating a cylinder (usually in a master-key system) to two or more different keys that normally would not be expected to operate together. See also *Controlled cross-keying* and *Uncontrolled cross-keying.*

CSI   Construction Specifiers Institute.

current   The flow of electricity. Current is measured in amperes.

cut   1. An indentation, notch, or cutout made in a key blank in order to make it operate a lock. See *Bitting.* 2. To make cuts into a key blade.

cut angle   A measurement, usually expressed in degrees, for the angle between the two sides of a key cut.

cut edge   A key that has been bitted or combinated.

cut root The bottom of a key cut.

cut-root shape The shape of the bottom of a key cut. It might have a flat or radius of a specific dimension or be a perfect V.

cutter The part of a key machine that makes the cuts into the key blank.

cylinder A complete lock operating unit that usually consists of the plug, shell, tumblers, springs, plug retainer, a cam/tailpiece, or other actuating device, and all other necessary operating parts.

cylinder blank A dummy cylinder that has a solid face and no operating parts.

cylinder clip A spring-steel device used to secure some types of cylinders.

cylinder collar A plate or ring installed under the head of a cylinder to improve appearance and/or security.

cylinder guard A protective cylinder mounting device.

cylinder key A broad generic term including virtually all pin- and disk-tumbler keys.

cylindrical lock See *Key-in-knob lock.*

cylindrical lockset A bored lockset whose latch or bolt locking mechanism is contained in the portion installed through the cross bore.

dc Direct current. Electric current that flows in one direction. For practical purposes, direct current is the type of electricity obtained from batteries.

deadbolt (or dead bolt) A lock bolt, usually rectangular, that has no spring action when in the locked position and that becomes locked against end pressure when fully projected.

deadbolt lock (or deadbolt), tubular deadbolt, tubular dead lock, cylindrical deadbolt A lock that projects a deadbolt.

deadlatch A lock with a beveled latchbolt that can be locked automatically or manually against end pressure when projected.

declining-step key A key whose cuts are progressively deeper from bow to tip.

decode To determine a key combination by physical measurement of a key and/or cylinder parts.

degree of rotation A specification for the angle at which a cut is made into a key blade as referenced from the perpendicular.

department master key A master key that operates all or most master-keyed locks of a given department.

depth key set A set of keys used to make a code-original key on a key duplicating machine to a lock manufacturer's given set of key bitting specifications. Each key is cut with the correct spacing to one depth only in all bitting positions, with one key for each depth.

derived series A series of blind codes and bittings that are directly related to those of another bitting list.

DHI Door and Hardware Institute.

dimple A key cut in a dimple key.

dimple key A key that has cuts drilled or milled into its blade surfaces. The cuts normally do not change the blade silhouette.

direct code   A designation assigned to a particular key that includes the actual combination of the key.

disk tumbler or wafer tumbler   1. A flat tumbler that must be drawn into the cylinder plug by the proper key so that none of its extremities extends into the shell. 2. A flat, usually rectangular tumbler with a gate that must be aligned with a sidebar by the proper key.

display key   A special change key in a hotel master-key system that will allow access to one designated guest room even if the lock is in the shut-out mode. It also might act as a shut-out key for that room.

DMM   Digital multimeter. A device used to measure current, resistance, and voltage.

double-acting hinge   A hinge that allows movement of a door in either direction from the closed position.

double-bitted key   A key bitted on two opposite surfaces.

double pin   To place more than one master pin in a single-pin chamber.

drive-in latch   A cylindrical latch with a cylindrical knurled ring around its face designed to grip the edge bore of a door. Drive-in lathes are used commonly with deadbolt locks on metal doors.

driver or top pin, upper pin   A cylindrical tumbler, usually flat on both ends, that is installed directly under the spring in the pin stack.

driver spring   A spring placed on top of the pin stack to exert pressure on the pin tumblers.

drop   A pivoting or swinging dust cover.

dummy cylinder   1. A nonfunctional facsimile of a rim or mortise cylinder used for appearance only, usually to conceal a cylinder hole. 2. A nonlocking device that looks like a cylinder and is used to cover up a cylinder hole.

duplicate key   Any key reproduced from a pattern key.

dust cover   A device designed to prevent foreign matter from entering a mechanism through the keyway.

dustproof cylinder   A cylinder designed to prevent foreign matter from entering either end of the keyway.

edge bore   A hole drilled into the edge of a door where a bored or interconnected lockset is to be installed.

effective plug diameter   The dimension obtained by adding the root depth of a key cut to the length of its corresponding bottom pin, which establishes a perfect shear line. This will not necessarily be the same as the actual plug diameter.

ejector hole   A hole found on the bottom of certain interchangeable cores under each pin chamber. It provides a path for the ejector pin.

ejector pin   A tool used to drive all the elements of a pin chamber out of certain interchangeable cores.

electric strike   An electrically controlled solenoid and mechanical latching device.

electrified lockset   A lock that is controlled electrically.

electromagnetic lock   A locking device that uses magnetism to keep it in a locked position.

emergency key   The key that operates a privacy-function lockset.

emergency master key   A special master key that usually operates all guest-room locks in a hotel master-key system at all times, even in the shut-out mode. This key also may act as a shut-out key.

emf   Electromotive force (also called *voltage*). The force needed to cause current to flow within a circuit.

EMK   Emergency master key.

ENG   Symbol for engineer's key.

escutcheon   A surface-mounted trim that enhances the appearance and/or security of a lock installation.

extractor key   A tool that normally removes a portion of a two-piece key or blocking device from a keyway.

face plate   A mortise-lock cover plate exposed in the edge of a door.

factory-original key   The cut key furnished by the lock manufacturer for a lock or cylinder.

fail-safe lock   A lock that automatically unlocks during a power failure.

fail-secure lock   A lock that automatically locks during a power failure.

fence   A projection on a lock bolt that prevents movement of the bolt unless it can enter gates of properly aligned tumblers.

finish   A material, coloring, and/or texturing specification.

fireman's key   A key used to override normal operation of elevators, bringing them to the ground floor.

first-generation duplicate   A key that was duplicated using a factory-original key or a code-original key as a pattern.

first key   Any key produced without the use of a pattern key.

five-column progression   A process wherein key bittings are obtained by using the cut possibilities in five columns of the key bitting array.

five-pin master key   A master key for all combinations obtained by progressing five bitting positions.

flexible-head mortise cylinder   An adjustable mortise cylinder that can be extended against spring pressure to a slightly longer length.

floor master key   A master key that operates all or most master keyed locks on a particular floor of a building.

following tool   See *Plug follower*.

four-column progression   A process wherein key bittings are obtained by using the cut possibilities in four columns of the key bitting array.

four-pin master key   A master key for all combinations obtained by progressing four bitting positions.

frangible shackle   A padlock shackle designed to be broken easily.

gate   A notch cut into the edge of a tumbler to accept a fence or sidebar.

graduated drivers   A set of top pins of different lengths. Use is based on the height of the rest of the pin stack in order to achieve a uniform pin stack height.

grand master key (GMK)   The key that operates two or more separate groups of locks, which are each operated by a different master key.

Grand-master-key system   A master-key system that has exactly three levels of keying.

grand master keyed   Of or pertaining to a lock or cylinder that is or is to be keyed into a grand-master-key system.

great grand master key (GGMK)   The key that operates two or more separate groups of locks, which are each operated by a different grand master key.

great-grand-master-key system   A master key system that has exactly four levels of keying.

great grand master keyed   Of or pertaining to a lock or cylinder that is or is to be keyed into a great-grand-master-key system.

great great grand master key (GGGMK)   The key that operates two or more separate groups of locks, which are each operated by different great grand master keys.

great-great-grand-master-key system   A master key system that has five or more levels of keying.

great great grand master keyed   Of or pertaining to a lock or cylinder that is or is to be keyed into a great-great-grand-master-key system.

ground   An electrical connection to a metallic object that is either buried in the earth or is connected to a metallic object buried in the earth.

guard key   A key that must be used in conjunction with a renter's key to unlock a safe deposit lock. It is usually the same for every lock within an installation.

guest key   A key in a hotel master-key system that is normally used to unlock only the one guest room for which it was intended but will not operate the lock in the shut-out mode.

guide   That part of a key machine that follows the cuts of a pattern key or template during duplication.

hardware schedule   A listing of the door hardware used on a particular job. It includes the types of hardware, manufacturers, locations, finishes, and sizes. It should include a keying schedule specifying how each locking device is to be keyed.

hasp   A hinged metal strap designed to be passed over a staple and secured in place.

heel & toe locking   Refers to a padlock that has locking dogs at both the heel and toe of the shackle.

heel (of a padlock shackle)   The part of a padlock shackle that is retained in the case.

high-security cylinder   A cylinder that offers a greater degree of resistance to any or all of the following: picking, impressioning, key duplication, drilling, or other forms of forcible entry.

high-security key   A key for a high-security cylinder.

hold-open cylinder   A cylinder provided with a special cam that will hold a latch bolt in the retracted position when so set by the key.

holding fixture   A device that holds cylinder plugs, cylinders, housings, and/or cores to facilitate the installation of tumblers, springs, and/or spring covers.

hollow driver   A top pin hollowed out on one end to receive the spring, typically used in cylinders with extremely limited clearance in the pin chambers.

horizontal group master (HGM) key   The two-pin master key for all combinations listed in all blocks in a line across the page in the standard progression format.

housekeeper's key (HKP)   A selective master key in a hotel master-key system that may operate all guest and linen rooms and other housekeeping areas.

housing   That part of a locking device designed to hold a core.

impression   1. The mark made by a tumbler on its key cut. 2. To fit a key by the impression technique.

IAHSSP   International Association of Home Safety and Security Professionals.

impression technique   A means of fitting a key directly to a locked cylinder by manipulating a blank in the keyway and cutting the blank where the tumblers have made marks.

incidental master key   A key cut to an unplanned shearline created when the cylinder is combinated to the top master key and a change key.

increment   A usually uniform increase or decrease in the successive depths of a key cut, which must be matched by a corresponding change in the tumblers.

indicator   A device that provides visual evidence that a deadbolt is extended or that a lock is in the shut-out mode.

individual key   An operating key for a lock or cylinder that is not part of a keying system.

insulator   Materials such as rubber and plastics that provide resistance to current flow. They are used to cover conductors and electrical devices to prevent unwanted current flow.

interchangeable core (IC)   A key-removable core that can be used in all or most of the core manufacturer's product line. No tools other than the control key are required for removal of the core.

interlocking pin tumbler   A type of pin tumbler that is designed to be linked together with all other tumblers in its chamber when the cylinder plug is in the locked position.

jamb   The vertical components of a door frame.

jimmy-proof   To provide a lock with a bolt that interlocks with its strike.

jumbo cylinder   A rim or mortise cylinder VA.

k   Symbol for keys used after a numerical designation of the quantity of the keys requested to be supplied with the cylinders: 1k, 2k, 3k, etc. It is usually found in hardware/keying schedules.

KA   Keyed alike. This symbol indicates cylinders that are to be operated by the same key(s)—for example: KA1, KA2, etc. KA/2, KA/3, etc. are the symbol used to indicate the quantity of locks or cylinders in keyed-alike groups. These groups are usually formed from a larger quantity.

KBA   Key bitting array.

KD   Keyed different.

keeper   See *Strike plate.*

key   A properly combined device that is or most closely resembles the device specifically intended by the lock manufacturer to operate the corresponding lock.

key bitting array   A matrix (graphic) of all possible bittings for change keys and master keys as related to the top master key.

key bitting punch   A manually operated device that stamps or punches the cuts into a key blade rather than grinding or milling them.

key bitting specifications   The technical data required to bit a given key blank or family of key blanks to the lock manufacturer's dimensions.

key blank   Any material manufactured to the proper size and configuration that allows its entry into the keyway of a specific locking device. A key blank that has not yet been combined or cut.

key changeable   Of or pertaining to a lock or cylinder that can be recombinated without disassembly by use of a key. The use of a tool also might be required.

key-coding machine   A key machine designed for the production of code keys. It might or might not also serve as a duplicating machine.

key control   1. Any method or procedure that limits unauthorized acquisition of a key and/or controls distribution of authorized keys. 2. A systematic organization of keys and key records.

key cut(s)   The portion of the key blade that remains after being cut and that aligns the tumbler(s).

key-cut profile   The shape of a key cut, including the cut angle and the cut root shape.

key-duplicating machine   A key machine that is designed to make copies from a pattern key.

key gauge   A device (usually flat) with a cutaway portion indexed with a given set of depth or spacing specifications. Used to help determine the combination of a key.

key-in-knob cylinder   A cylinder used in a key-in-knob lock.

key-in-knob lock or cylindrical lock   A lock that uses one or two knobs and has a lock cylinder in one or both knobs.

key interchange   An undesirable condition, usually in a master-key system, whereby a key unintentionally operates a cylinder or lock.

key machine   Any machine designed to cut keys.

key manipulation   Manipulation of an incorrect key in order to operate a lock or cylinder.

key milling   The grooves machined into the length of a key blade to allow its entry into the keyway.

key override   A provision allowing interruption or circumvention of normal operation of a combination lock or electrical device.

key-override cylinder   A lock cylinder installed in a device to provide a key-override function.

key pull position   Any position of the cylinder plug at which the key can be removed.

key records   Records that typically include some or all of the following: bitting list, key bitting array, key-system schematic, end user, number of keys/cylinders issued, names of persons to whom keys were issued, hardware/keying schedule.

key retaining   1. Of or pertaining to a lock that must be locked before its key can be removed. 2. Of or pertaining to a cylinder or lock that may prevent removal of a key without the use of an additional key and/or tool.

key section   The exact cross-sectional configuration of a key blade as viewed from the bow toward the tip.

keyswitch   A switch that is operated with a key.

key-system schematic   A drawing with blocks using keying symbols, usually illustrating the hierarchy of all keys within a master-key system. It indicates the structure and total expansion of the system.

key-trap core/cylinder   A special core or cylinder designed to capture any key to which it is combined once that key is inserted and turned slightly.

keyed   1. Combined. 2. Having provision for operation by key.

keyed alike   Of or pertaining to two or more locks or cylinders that have the same combination. They may or may not be part of a keying system.

keyed random   Of or pertaining to a cylinder or group of cylinders selected from a limited inventory of different key changes. Duplicate bittings may occur.

keying   Any specification for how a cylinder or group of cylinders are combined in order to control access.

keying conference   A meeting of the end user and the keying-system supplier at which the keying and levels of keying, including future expansion, are determined and specified.

keying kit   A compartmented container that holds an assortment of tumblers, springs, and/or other parts.

keying schedule   A detailed specification of the keying system listing how all cylinders are to be keyed and the quantities, markings, and shipping instructions of all keys and/or other parts.

keying symbol   A designation used for a lock or cylinder combination in the standard key coding system: AA1, XAA1, XIX, etc.

keyway   1. The opening in a lock or cylinder shaped to accept a key bit or blade of a proper configuration. 2. The exact cross-sectional configuration of a keyway as viewed from the front. It is not necessarily the same as the key section.

keyway unit   The plug of certain binary-type disk-tumbler key-in-knob locks.

KR   1. Keyed random. 2. Key retaining.

KWY   Keyway.

layout tray   A compartmented container used to organize cylinder parts during keying or servicing.

lazy cam/tailpiece   A cam or tailpiece designed to remain stationary while the cylinder plug is partially rotated (or vice versa).

LCD   Liquid-crystal display.

lever lock or key-in-lever lock, lever-handle lock   A lock that has a lever handle.

lever tumbler   A flat, spring-loaded tumbler that pivots on a post. It contains a gate that must be aligned with a fence to allow movement of the bolt.

Lever-tumbler lock or lever lock   A lock that has a lever-tumbler cylinder.

loading tool   A tool that aids installation of cylinder components into the cylinder shell.

lock   A device that incorporates a bolt, cam, shackle, or switch to secure an object—such as a door, drawer, or machine—to a closed, locked, off or on position and that provides a restricted means of releasing the object from that position.

Lock-plate compressor or lock-plate remover   A tool designed to depress the lock plate of an automobile steering column. The tool is used when disassembling automobile steering columns.

lockout   Any situation in which the normal operation of a lock or cylinder is prevented.

lockout key   A key made in two pieces. One piece is trapped in the keyway by the tumblers when inserted and blocks entry of any regular key. The second piece is used to remove the first piece.

lockset   A locking device complete with trim mounting hardware and strike.

locksmith   A person trained to install, service, and bypass locking devices.

mA   Milliampere.

MACS   Maximum adjacent cut specification.

maid's master key   The master key in a hotel master-key system given to the maid. It operates only cylinders of the guest rooms and linen closets in the maid's designated area.

maison key system   A keying system in which one or more cylinders are operated by every key (or relatively large numbers of different keys) in the system—for example, main entrances of apartment buildings operated by all individual suite keys of the building.

manipulation key   Any key other than a correct key that can be variably positioned and/or manipulated in a keyway to operate a lock or cylinder.

master disk   A special disk tumbler with multiple gates to receive a sidebar.

master key (MK)   1. A key that operates all the master-keyed locks or cylinders in a group, with each lock or cylinder operated by its own change key. 2. To combinate a group of locks or cylinders such that each is operated by its own change key as well as by a master key for the entire group.

master-key changes   The number of different usable change keys available under a given master key.

master-key system   1. Any keying arrangement that has two or more levels of keying. 2. A keying arrangement that has exactly two levels of keying.

master keyed   Of or pertaining to a cylinder or group of cylinders that are combined so that all may be operated by their own change keys and by additional keys known as master keys.

master keyed only   Of or pertaining to a lock or cylinder that is combinated only to a master key.

master lever   A lever tumbler that can align some or all other levers in its lock so that lever gates are at the fence. It is typically used in locker locks.

master pin   1. Usually a cylindrical tumbler, flat on both ends, placed between the top and bottom pins to create an additional shear line. *2.* A pin tumbler with multiple gates to accept a sidebar.

master ring   A tube-shaped sleeve located between the plug and shell of certain cylinders to create a second shear line. Normally, the plug shear line is used for change-key combinations, and the shell shear line is used for master-key combinations.

master-ring lock/cylinder   A lock or cylinder equipped with a master ring.

master wafer   A ward used in certain binary-type disk-tumbler key-in-knob locks.

maximum adjacent cut specification   The maximum allowable depths to which opposing cuts can be made without breaking through the key blade. This is typically a consideration with dimple keys.

metal oxide varister   A voltage-dependent resistor.

miscut   Of or pertaining to a key that has been cut incorrectly.

MOCS   Maximum opposing cut specification.

mogul cylinder   A very large pin-tumbler cylinder whose pins, springs, key, etc. are also proportionately increased in size. It is frequently used in prison locks.

mortise   An opening made in a door to receive a lock or other hardware.

mortise cylinder   A threaded cylinder typically used in mortise locks of American manufacture.

MOV   Metal oxide varister.

mullion   A vertical center post in the frame of a pair of doors.

multisection key blank   A key section that enters more than one, but not all, keyways in a multiplex key system.

multiple gating   A means of master keying by providing a tumbler with more than one gate.

multiplex key blank   Any key blank that is part of a multiplex key system.

multiplex key system   1. A series of different key sections that may be used to expand a master-key system by repeating bittings on additional key sections. The keys of one section will not enter the keyway of another key section. This type of system always includes another key section that will enter more than one or all of the keyways. 2. A keying system that uses such keyways and key sections.

multitester   A device designed to measure current, resistance, and voltage. The VOM and the DMM are two types of multitesters.

mushroom pin   A pin tumbler, usually a top pin, that resembles a mushroom. It is typically used to increase pick resistance.

NC (or N/C)   Normally closed.

NCK   No change key. This symbol is used primarily in hardware schedules.

negative locking   Locking achieved solely by spring pressure or gravity, which prevents a key cut too deeply from operating a lock cylinder.

nickel-cadmium   A long-life rechargeable battery or cell often used as a backup power supply.

night latch   An auxiliary rim lock with a spring latchbolt.

NMK   Not master keyed. This keying symbol is suffixed in parentheses to the regular key symbol. It indicates that the cylinder is not to be operated by the master key(s) specified in the regular key symbol—for example: AB6(NMK).

NO (or N/O)   Normally open.

nonkey-retaining (NKR)   Of or pertaining to a lock whose key can be removed in both the locked and unlocked positions.

nonkeyed   Having no provision for key operation. *Note*: This term also includes privacy-function locksets operated by an emergency key.

nonoriginal key blank   Any key blank other than an original.

nonremovable pin   A pin that cannot be removed from a hinge when the door is closed.

normally closed switch   A switch whose contacts normally remain closed when electric current is not flowing through it.

normally open switch   A switch whose contacts normally remain open when electric current is not flowing through it.

NSLA   National Locksmith Suppliers Association.

NSO   National Safeman's Organization.

odometer method   A means of progressing key bittings using a progression sequence of right to left.

ohm   A unit of measure of resistance to electrical current flow.

Ohm's law   The description of the relationship between voltage, current, and resistance in an electrical circuit. Ohm's law states that a resistance of one ohm passes through a current of one ampere in response to one volt. Mathematically expressed, $E = IR$, with E as voltage in volts, I as current in amperes, and R as resistance in ohms.

one-bitted   Of or pertaining to a cylinder that is combinated to keys cut to the manufacturer's reference number one bitting.

one-column progression   A process wherein key bittings are obtained by using the cut possibilities in one column of the bitting array.

one-pin master key   A master key for all combinations obtained by progressing only one bitting position.

operating key   Any key that will properly operate a lock or cylinder to lock or unlock the lock mechanism and is not a control key or reset key.

operating shear line   Any shear line that allows normal operation of a cylinder or lock.

original key blank   A key blank supplied by the lock manufacturer to fit that manufacturer's specific product.

padlock   A detachable and portable lock with a shackle that locks into its case.

page master key   The three-pin master key for all combinations listed on a page in the standard progression format.

panic bar   A door-mounted exit bar designed to allow fast egress from the inside of the door and resistance to entry from the outside the door.

paracentric   Of or pertaining to a keyway with one or more wards on each side projecting beyond the vertical center line of the keyway to hinder picking.

parallel circuit   An electrical circuit that provides more than one path for current to flow.

PASS-Key   Personalized Automotive Security System. See *VATS*.

pass key   A master key or skeleton key.

pattern key   1. A key kept on file to use in a key-duplicating machine when additional keys are required. 2. Any key used in a key-duplicating machine to create a duplicate key.

pawl   A tailpiece or cam for an automobile door or deck lock.

pick   1. A tool or instrument, other than the specifically designed key, made for the purpose of manipulating tumblers in a lock or cylinder into the locked or unlocked position through the keyway without obvious damage. 2. To manipulate tumblers in a keyed-lock mechanism through the keyway without obvious damage by means other than the specifically designed key.

pick key   A type of manipulation key cut or modified to operate a lock or cylinder.

pin   To install pin tumblers into a cylinder and/or cylinder plug.

pin chamber   The corresponding hole drilled into the cylinder shell and/or plug to accept the pin(s) and spring.

pin kit   A type of keying kit for a pin-tumbler mechanism.

pin stack   All the tumblers in a given pin chamber.

pin-stack height   The measurement of a pin stack, often expressed in units of the lock manufacturer's increment or as an actual dimension.

pin tumbler   Usually a cylindrical-shaped tumbler. Three types are normally used: bottom pin, master pin, and top pin.

pin tweezers   A tool used in handling tumblers and springs.

pinning block   A holding fixture that assists in the loading of tumblers into a cylinder or cylinder plug.

pinning chart   A numerical diagram that indicates the sizes and order of installation of the various pins into a cylinder. The sizes usually are indicated by a manufacturer's reference number, which equals the quantity of increments a tumbler represents.

plug   The part of a cylinder containing the keyway, with tumbler chambers usually corresponding to those in the cylinder shell.

plug follower   A tool used to allow removal of the cylinder plug while retaining the top pins, springs, and/or other components within the shell.

plug holder   A holding fixture that assists in the loading of tumblers into a cylinder plug.

plug retainer   The cylinder component that secures the plug in the shell.

plug spinner   A tool designed to quickly spin a plug clockwise or counterclockwise into an unlocked position, when the lock has been picked, into a position that doesn't allow the lock to open.

positional master keying   A method of master keying typical of certain binary-type disk-tumbler key-in-knob locks and of magnetic and dimple-key cylinders. Of all possible tumbler positions within a cylinder, only a limited number contain active tumblers. The locations of these active tumblers are rotated among all possible positions to generate key changes. Higher-level keys must have more cuts or magnets than lower-level keys.

positive locking   The condition brought about when a key cut that is too high forces its tumbler into the locking position. This type of locking does not rely on gravity or spring pressure.

post (of a key)   The portion of a bit key between the tip and the shoulder to which the bit or bits are attached, **practical key changes**   The total number of usable different combinations available for a specific cylinder or lock mechanism.

prep key   A type of guard key for a safe deposit box lock with only one keyway. It must be turned once and withdrawn before the renter's key will unlock the unit.

privacy key   A key that operates an SKD cylinder.

profile cylinder   A cylinder with a uniform cross section that slides into place and is held by a mounting screw. It is typically used in mortise locks of non-U.S. manufacturers.

progress   To select possible key bittings, usually in numerical order, from the key-bitting array.

progression   A logical sequence of selecting possible key bittings, usually in numerical order from the key-bitting array.

progression column   A listing of the key-bitting possibilities available in one bitting position as displayed in a column of the key-bitting array.

progression list   A bitting list of change keys and master keys arranged in sequence of progression.

progressive   Any bitting position that is progressed rather than held constant.

proprietary   Of or pertaining to a keyway and key section assigned exclusively to one end user by the lock manufacturer. It also may be protected by law from duplication.

radiused blade bottom   The bottom of a key blade that has been radiused to conform to the curvature of the cylinder plug it is designed to enter.

random master keying   Any undesirable process of master keying that uses unrelated keys to create a system.

rap   1. To unlock a plug from its shell by striking sharp blows to the spring side of the cylinder while applying tension to the plug. 2. To unlock a padlock shackle from its case by striking sharp blows to the sides in order to disengage the locking dogs.

read key   A key that allows access to the sales and/or customer data on certain types of cash control equipment, such as a cash register.

recombinate   To change the combination of a lock, cylinder, or key.

recore   To rekey by installing a different core.

register groove   The reference point on the key blade from which some manufacturers locate the bitting depths.

register number   1. A reference number typically assigned by the lock manufacturer to an entire master-key system. 2. A blind code assigned by some lock manufacturers to higher-level keys in a master-key system.

rekey (or re-key)   To change the existing combination of a cylinder or lock so that the lock can be operated by a different key.

relay   A type of switching device, usually electronic or electromechanical.

removable core   A key-removable core that can only be installed in one type of cylinder housing—for example, rim cylinder or mortise cylinder or key-in-knob locks.

removable cylinder   A cylinder that can be removed from a locking device by a key and/or tool.

removal key   The part of a two-piece key that is used to remove its counterpart from a keyway.

renter's key   A key that must be used together with a guard key, prep key, or electronic release to unlock a safe deposit lock. It is usually different for every unit within an installation.

repin   To replace pin tumblers, with or without changing the existing combination.

reset key   1. A key used to set some types of cylinders to a new combination. Many of these cylinders require the additional use of tools and/or the new operating key to establish the new combination. 2. A key that allows the tabulations on various types of cash-control equipment such as cash registers to be cleared from the records of the equipment.

resistance   Opposition to electric current flow.

resistor   A component that resists electric current flow in a dc circuit.

restricted   Of or pertaining to a keyway and corresponding key blank whose sale and/or distribution is limited by the lock manufacturer in order to reduce unauthorized key proliferation.

retainer-clip tool   A tool designed to install or remove retainer clips from automobiles.

reversible key   A symmetric key that may be inserted either up or down to operate a lock.

reversible lock   A lock in which the latch bolt can be turned over and adapted to doors of either hand, opening in or out.

rim cylinder   A cylinder typically used with surface-applied locks and attached with a back plate and machine screws. It has a tailpiece to actuate the lock mechanism.

RL   Registered Locksmith, as certified by Associated Locksmiths of America.

RM   Row master key.

root depth   The dimension from the bottom of a cut on a key to the bottom of the blade.

rose   A usually circular escutcheon.

rotary tumbler   A circular tumbler with one or more gates. Rotation of the proper key aligns the tumbler gates at a sidebar, fence, or shackle slot.

rotating constant   One or more cut(s) in a key of any level that remain constant throughout all levels and are identical to the top master-key cuts in their corresponding positions. The positions where the top master-key cuts are held constant may be moved, always in a logical sequence.

rotating constant method   A method used to progress key bittings in a master-key system wherein at least one cut in each key is identical to the corresponding cut in the top master key. The identical cut is moved to a different location in a logical sequence until each possible planned position has been used.

row master key   The one-pin master key for all combinations listed on the same line across the page in the standard progression format.

RPL   Registered Professional Locksmith, as certified by the International Association of Home Safety and Security Professionals.

RSP   Registered Security Professional, as certified by the International Association of Home Safety and Security Professionals.

RST   Registered Safe Technician, as certified by the National Safeman's Organization.

S/A   Subassembled.

SAVTA   Safe and Vault Technicians Association.

scalp   A thin piece of metal that is usually crimped or spun onto the front of a cylinder. It determines the cylinder's finish and also may serve as the plug retainer.

second-generation duplicate   A key reproduced from a first-generation duplicate.

security collar   A protective cylinder collar.

segmented follower   A plug follower that is sliced into sections, which are introduced into the cylinder shell one at a time. It is typically used with profile cylinders.

selective key system   A key system in which every key has the capacity of being a master key. It is normally used for applications requiring a limited number of keys and extensive cross-keying.

selective master key   An unassociated master key that can be made to operate any specific lock in the entire system, in addition to the regular master key and/or change key for the cylinder, without creating key interchange.

sequence of progression   The order in which bitting positions are progressed to obtain change-key combinations.

series circuit   An electrical circuit that provides only one path for current flow.

series wafer   A type of disk tumbler used in certain binary-type disk-tumbler key-in-knob locks. Its presence requires that no cut be made in that position on the operating key(s).

setup key   A key used to calibrate some types of key machines.

setup plug   A type of loading tool shaped like a plug follower. It contains pin chambers and is used with a shove knife to load springs and top pins into a cylinder shell.

seven-column progression   A process wherein key bittings are obtained by using the cut possibilities in seven columns of the key-bitting array.

seven-pin master key   A master key for all combinations obtained by progressing seven bitting positions.

shackle   The usually curved portion of a padlock that passes though a hasp and snaps into the padlock's body.

shackle spring   The spring inside the body of a padlock that allows the shackle to pop out of the body when in the unlocked position.

shank   The part of a bit key between the shoulder and the bow.

shear line   A location in a cylinder at which specific tumbler surfaces must be aligned, removing obstruction(s) that prevent the plug from moving.

shell   The part of the cylinder that surrounds the plug and that usually contains tumbler chambers corresponding to those in the plug.

shim   1. A thin piece of material used to unlock the cylinder plug from the shell by separating the pin tumblers at the shear line one at a time. 2. To unlock a cylinder plug from its shell using a shim.

shoulder   Any key stop other than a tip stop.

shouldered pin   A bottom pin whose diameter is larger at the flat end to limit its penetration into a counterbored chamber.

shove knife   A tool used with a setup plug that pushes the springs and pin tumblers into the cylinder shell.

shut-out key   Usually used in hotel keying systems, a key that will make the lock inoperative to all other keys in the system except the emergency master key, display key, and some types of shut-out keys.

shut-out mode   The state of a hotel-function lockset that prevents operation by all keys except the emergency master key, display key, and some types of shut-out keys.

sidebar   A primary or secondary locking device in a cylinder. When locked, it extends along the plug beyond its circumference. It must enter gates in the tumblers in order to clear the shell and allow the plug to rotate.

simplex-key section   A single independent key section that cannot be used in a multiplex key section.

single-key section   An individual key section that can be used in a multiplex key system.

single-step progression   A progression using a one increment difference between bittings of a given position.

six-column progression   A process wherein key bittings are obtained by using the cut possibilities in six columns of the key-bitting array.

six-pin master key   A master key for all combinations obtained by progressing six bitting positions.

SKD   Symbol for single keyed. Normally followed by a numerical designation in the standard key coding system—SKD1, SKD2, etc. It indicates that a cylinder or lock is not master keyed but is part of the keying system.

skeleton key   A warded lock key cut especially thin to bypass the wards in several warded locks so that the locks can be opened. The term is used commonly by laypersons to refer to any bit key.

spacing   The dimensions from the stop to the center of the first cut and/or to the centers of successive cuts.

special-application cylinder   Any cylinder other than a mortise, rim, key-in-knob, or profile cylinder.

split-pin master keying   A method of master keying a pin-tumbler cylinder by installing master pins into one or more pin chambers.

spool pin   Usually a top pin that resembles a spool, typically used to increase pick resistance.

spring cover   A device for sealing one or more pin chambers.

standard key-coding system   An industry standard and uniform method of designating all keys and/or cylinders in a master-key system. The designation automatically indicates the exact function and keying level of each key and/or cylinder in the system, usually without further explanation.

standard progression format   A systematic method of listing and relating all change-key combinations in a master-key system. The listing is divided into segments known as blocks, horizontal groups, vertical groups, rows, and pages for levels of control.

step pin   A spool or mushroom pin that has had a portion of its end machined to a smaller diameter than the opposite end. It is typically used as a top pin to improve pick resistance by some manufacturers of high-security cylinders.

stepped tumbler   A special (usually disk) tumbler used in master keying. It has multiple bearing surfaces for blades of different key sections.

stop (of a key)   The part of a key from which all cuts are indexed and which determines how far the key enters the keyway.

strike   See *Strike plate* and *Strike box.*

strike box or box strike   A strike plate that incorporates a metal or plastic box that encloses the lock's bolt or latch when in a locked position.

strike plate or strike, keeper   The part of a locking arrangement that receives the bolt, latch, or fastener when the lock is in the locked position. The strike is usually recessed in a door frame.

sub-master key (SMK)   The master-key level immediately below the master key in a system of six or more levels of keying.

switch   A device used for opening and closing a circuit.

tailpiece   An actuator attached to the rear of the cylinder, parallel to the plug, typically used on rim, key-in-knob, or special-application cylinders.

tension wrench or turning tool, torque wrench, torsion wrench, turning wrench   A tool, usually made of spring steel, used in conjunction with a lock pick. While the pick is moving a lock's tumblers into a position that frees the plug to be rotated, the tension wrench places rotational force on the plug. Although *tension wrench* is older and used more commonly than any of its synonyms, a large minority of locksmiths think the term is technically inaccurate and should not be used by locksmiths.

theoretical key changes   The total possible number of different combinations available for a specific cylinder or lock mechanism.

three-column progression   A process wherein key bittings are obtained by using the cut possibilities in three columns of the key-bitting array.

three-pin master key   A master key for all combinations obtained by progressing three bitting positions.

thumb-turn cylinder   A cylinder with a turn knob rather than a keyway and tumbler mechanism.

tip   The portion of the key that enters the keyway first.

tip stop   A type of stop located at or near the tip of the key.

tolerance   The deviation allowed from a given dimension.

top master key (TMK)   The highest-level master key in a master-key system.

top of blade   The bitted edge of a single-bitted key.

top pin   See *Driver.*

torque wrench   See *Tension wrench.*

torsion wrench   See *Tension wrench.*

total position progression   A process used to obtain key bittings in a master-key system wherein bittings of change keys differ from those of the top master key in all bitting positions.

transformer   A device that transfers electrical energy from one circuit to another without direct connection between them.

try-out key   A manipulation key that is usually part of a set, used for a specific series, keyway, and/or brand of lock.

tubular key   A key with a tubular blade. The key cuts are made into the end of the blade, around the circumference.

Tubular-key lock   A type of lock with tumblers arranged in a circle, often used on vending machines and coin-operated washing machines.

tubular lock or key-in-knob lock, tubular key-in-knob lock   A key-in-knob lock that has two screw posts, one on each side of the lock's central spindle.

tumbler   A movable obstruction of varying size and configuration in a lock or cylinder that makes direct contact with the key or another tumbler and prevents an incorrect key or torquing device from activating the lock or other mechanism.

tumbler spring   Any spring that acts directly on a tumbler.

turning tool   See *Tension wrench.*

turning wrench   See *Tension wrench.*

two-column progression   A process wherein key bittings are obtained by using the cut possibilities in two columns of the key-bitting array.

two-pin master key   A master key for all combinations obtained by progressing two bitting positions.

two-step progression   A progression using a two-increment difference between bittings of a given position.

UL   Underwriters Laboratories.

unassociated change key   A change key that is not related directly to a particular master key through the use of certain constant cuts.

unassociated master key   A master key that does not have change keys related to its combination through the use of constant cuts.

uncombinated   1. Of or pertaining to a cylinder that is supplied without keys, tumblers, and springs. 2. Of or pertaining to a lock, cylinder, or key in which the combination has not been set.

uncontrolled cross-keying   A condition in which two or more different keys under different higher-level keys operate one cylinder by design—for example: XAA1 operated by AB, AB1. *Note*: This condition severely limits the security of the cylinder and the maximum expansion of the system and often leads to key interchange.

unidirectional cylinder   A cylinder whose key can turn in only one direction from the key-pull position, often not making a complete rotation.

upper pin   See *Driver*.

Vac   ac volts.

VATS   Vehicle Anti-Theft System.

VATS decoder   A device designed for determining which VATS key blank to use to duplicate a VATS key or to make a VATS first key.

VATS key   A key designed to operate a vehicle equipped with VATS.

Vdc   dc volts.

vehicle antitheft system   An electromechanical system used in many General Motors vehicles to deter theft. Sometimes referred to as the PASS-Key system.

vertical-group master key (VGM)   The two-pin master key for all combinations listed in all blocks in a line down a page in the standard progression format.

visual key control (VKC)   A specification that all keys and the visible portion of the front of all lock cylinders be stamped with standard keying symbols.

volt   A unit of measure for voltage.

voltage   The force that pushes electric current. Voltage (or electromotive force) is measured in volts.

voltage drop   The change in voltage across an electrical component (such as a resistor) in a circuit.

vom   Volt-ohm-milliammeter. A type of multitester.

ward   A usually stationary obstruction in a lock or cylinder that prevents the entry and/or operation of an incorrect key.

ward cut   A modification of a key that allows it to bypass a ward.

warded lock   A lock that relies primarily on wards to prevent unauthorized keys from operating the lock.

watt   A unit of measure of electrical power.

X   Symbol used in hardware schedules to indicate a cross-keyed condition for a particular cylinder—for example, XAA2, XlX (but not AX7).

zero bitted   Of or pertaining to a cylinder that is combinated to keys cut to the manufacturer's reference number 0 bitting.

Index

A-lamp, 121
Abloy Disklock, 67
Access-control systems, 111–119
    apartments, 113
    biometric system, 111–117
    card systems, 111
    password, 115–116
    Trilogy pushbutton lock, 118–119
Ace lock, 14
Acrylics, 59
Active alarm, 144
ActiveX, 176, 178
Adjust-a-Strike, 30
ADT Security Services, 15–16
Aftermarket blanks, 76
Airphone hands-free color video master
    station, 130, 131
Alarm installer, 183–184
Alarm Lock Model 250, 46–47
Alarm Lock Model 250L, 46–47
Alarm Lock Model 260, 46–47
Alarm Lock Model 260L, 46–47
Alarm Lock Model 265, 47–50
Alarm Lock Model 700, 50–53
Alarm Lock Model 700L, 50–53
Alarm Lock Model 710, 50–53
Alarm Lock Model 710L, 50–53
Alarm Lock Model 715, 53–56
Alarm system control panel, 98
Alarm system window stickers, 98, 104
Alarms, 97–110. *See also* Burglar alarms;
    Car alarms
All-risk insurance policy, 152
Allenbaugh, Howard, 27
Allenbaugh, Mark H., 27, 30
ALOA, 179
American Society for Industrial Security
    (ASIS), 185
Ancient Egyptian lock, 2, 3
Ancient Greece, 3–4

Ancient Rome, 5
Andrews, Solomon, 11
ANSI codes, 244
ANSI-graded locks, 64
Anti-Pinkertonism laws, 16
Antilift plates, 25
Antilift screws, 25
Antiscanning circuitry, 146
Antispyware, 176–177
Antivirus program, 179
Apartment security, 167–168
Applied biometric, 115
Architectural finish codes, 243–245
Art of the Locksmith, The, 7
ASIS, 185
Associated Locksmiths of America
    (ALOA), 179
Astor, Saul D., 1
Attack test, 76
Audio discriminators, 100
Augustus, Caesar, 5
Automobile security, 143–150
    car alarm, 144–148
    carjacking, 149–150
    cutoff switches, 143
    steering wheel lock, 143–144
    stolen-vehicle retrieval system, 148–149
    supplemental lock, 143–144
Axial-load test, 38

Backdoor, 177
Barron, Robert, 7
Best, Frank E., 13
BHMA codes, 244–245
Biaxial lock, 18, 85–86
Biometric authentication, 115
Biometric record, 116
Biometric system, 111–117
Bit-key lock, 64, 65
Black-and-white camera, 128

ABOUT THE AUTHOR

BILL PHILLIPS is president of the International Association of Home Safety and Security Professionals. He is also the most successful author of security books ever. His best-selling *The Complete Book of Locks and Locksmithing* is currently in its sixth edition. Mr. Phillips also wrote McGraw-Hill's *Locksmithing*, part of the Craftmaster Series, and *The Complete Book of Electronic Security*, also from McGraw-Hill. He is the author of the "Lock" article in the 1998 to 2006 editions of the *World Book Encyclopedia*, and has written hundreds of security-related articles for professional and general-circulation periodicals, including *Keynotes, Home Mechanix* (security editor), *Locksmith Ledger International* (contributing editor), *Los Angeles Times, Consumers Digest,* the *National Locksmith* (contributing writer), *Safe and Vault Technology, Security Dealer*, and many others.